SpringerBriefs in Applied Sciences and Technology

SpringerBriefs present concise summaries of cutting-edge research and practical applications across a wide spectrum of fields. Featuring compact volumes of 50 to 125 pages, the series covers a range of content from professional to academic.

Typical publications can be:

- A timely report of state-of-the art methods
- An introduction to or a manual for the application of mathematical or computer techniques
- A bridge between new research results, as published in journal articles
- A snapshot of a hot or emerging topic
- An in-depth case study
- A presentation of core concepts that students must understand in order to make independent contributions

SpringerBriefs are characterized by fast, global electronic dissemination, standard publishing contracts, standardized manuscript preparation and formatting guidelines, and expedited production schedules.

On the one hand, **SpringerBriefs in Applied Sciences and Technology** are devoted to the publication of fundamentals and applications within the different classical engineering disciplines as well as in interdisciplinary fields that recently emerged between these areas. On the other hand, as the boundary separating fundamental research and applied technology is more and more dissolving, this series is particularly open to trans-disciplinary topics between fundamental science and engineering.

Indexed by EI-Compendex, SCOPUS and Springerlink.

More information about this series at http://www.springer.com/series/8884

Tarek Al-Arbi Omar Ganat

Technical Guidance for Petroleum Exploration and Production Plans

 Springer

Tarek Al-Arbi Omar Ganat (iD)
Department of Petroleum Engineering
Universiti Teknologi Petronas
Seri Iskender, Malaysia

ISSN 2191-530X   ISSN 2191-5318   (electronic)
SpringerBriefs in Applied Sciences and Technology
ISBN 978-3-030-45249-0   ISBN 978-3-030-45250-6   (eBook)
https://doi.org/10.1007/978-3-030-45250-6

This Springer imprint is published by the registered company Springer Nature Switzerland AG
The registered company address is: Gewerbestrasse 11, 6330 Cham, Switzerland

I would like to dedicate this book to my parents and my brothers and sisters, and to my wife Basma and my children Mohamed, Heba, Abdulrahman, and my young hero Abdul Malik. Without their support and encouragement, this book will not complete.

Preface

This book was written after long of 25 years' experience in oil and gas industry, to cover the field development plan process and evaluation procedures due to a lack of literature in terms of real technical experiences to develop oil and gas projects. Based on my work experience as reservoir engineer in multi-regions located at different continentals, I have got much of knowledge and skill which encouraged me to share with others to make their evaluation process more easer.

The book provides a quick evaluation for any oil or gas discovery by using analogy information and experience knowledge (rule of thumb) to screen any available chances in exploration areas in the absence of many technical data.

Finally, I hope my book will be very useful for the petroleum engineers and can be used as good reference and a guidance, to make the discovery evaluation better and with less uncertainty.

Seri Iskender, Malaysia

Tarek Al-Arbi Omar Ganat

Overview of Chapters

This book covering several technical information's presented in the nine chapters. All the chapters defining their content; by offering the reader a complete view of the oil and gas field development plan. This will help the petroleum/reservoir engineers to understand the evaluation, the assessment process and the procedures required during the appraisal of oil and gas discoveries.

The chapters were scribed in a simple way to serve as a quick reference and guidance based on the discovery type and the amount of reservoir rock and fluid data available. Moreover, there is a strong and logical connection between all the chapters, where every chapter was written and in a structured way to achieve the milestone of this book. The arrangement of the chapters; lead to the full understanding of the field development plan process starting from hydrocarbon exploration through to economic analysis.

The book includes, real case studies were presented to demonstrate the stages covered in this book, facilitating petroleum/reservoir engineers to grasp the process of evaluation of the whole project stages. Deepwater projects for oil and gas prospects were presented and show in details the step by step evaluation process.

Contents

Chapter 1
Hydrocarbon Exploration

Hydrocarbon (Oil and gas) exploration is the search by geologists and geophysicists for oil and natural gas deposits underground for millions of years. The hydrocarbon exploration is congregated under the skill of petroleum geology.

1.1 Oil and Gas Estimation Reserves

Most significant factor in planning the field development and production of a reservoir oil or gas is a dependable estimate of the oil and gas volume in place (the natural resources), and estimate the expected oil and gas amount that can be recovered. This can be estimated at early stage where there are no much data available on the reservoir behaviour and production conditions. However, estimating hydrocarbon reserves is a complex practice that includes integrating geological and engineering data. Depending on the quantity and quality of data obtainable, one or more of the following direct or indirect methods could be used to estimate reserves:

1. Volumetric method delivers a static measure of oil or gas in place. This method uses at early in life of field. The accuracy of volumetric depends on data available such as porosity, net thickness, areal extent, hydrocarbon saturations.
2. The material balance is a subset of the mathematical techniques that are used by petroleum engineers. This technique, mathematically models the reservoir as a tank. The method uses some assumptions and tries to equilibrate changes in reservoir fluid volume as a result of production process. This method can be accounted aquifer support and gas cap expansion. Therefore, material balance method delivers a dynamic measure of hydrocarbon volumes at late stage of the production life of the field (assumes adequate production history available). Use in a mature field with abundant geological, petrophysical, and engineering data. The accuracy depends on quality of the input data such as pressure and temperature measurement surveys, and analysis of recovered fluids. The evaluation of the

T. A. O. Ganat, *Technical Guidance for Petroleum Exploration and Production Plans*, SpringerBriefs in Applied Sciences and Technology, https://doi.org/10.1007/978-3-030-45250-6_1

initial volumetric reserves can be verified by material balance calculations based on the production history of the oil field. Generally, the longer the production history data, the better the estimate reserve will be.

3. Recoverable reserves method, use after a reasonable amount of production data is available. Accuracy dependent on amount of production history data available. The reserve estimates tend to be realistic.

4. Analogy method, use at early stage of exploration and initial field development in order to estimate the hydrocarbon recovery. Estimating recovery in this way is particularly valuable when some production performance history is available but the production decline rate has not start yet. This method should always be used together with other techniques to make sure that the results make sense within the geological structure. Analogy is extremely dependent on similarity of reservoir characteristics. The reserve estimates are often very general. Therefore, the analogy method directly compares a newly discovered or poorly defined reservoir to a known reservoir that have similar geological depositional environment or reservoir rock and fluid properties.

Generally, the technique of estimating oil and gas reserves for a producing oil and gas fields is carry on through the life of the oil and gas fields. Always there is uncertainty to estimates oil and gas reserves. Actually, the degree of uncertainty is affected by any of the following factors:

1. Reservoir type,
2. Type and source of reservoir energy,
3. Type of quantity and quality of the geological, petrophysical, and engineering data,
4. Assumptions used for making the estimations,
5. Type of available technology, and
6. Skill and knowledge of the assessor.

However, the uncertainty decreases with time until the production rate reaches the economic limit and the ultimate recovery is reached, see Fig. 1.1.

When calculating reserves using any of the above methods, two calculation techniques could be used: deterministic and/or probabilistic.

The deterministic method is definitely the most common used technique. Deterministic methods derive proved reserves that are more tangible and understandable. In this method the parameters used in the calculation are exactly identified. The process is to select a single value for each parameter to input into the right equation, to obtain a single answer.

The probabilistic method, instead, is more rigorous and less generally used. This method uses a distribution curve for each single parameter using Monte Carlo Simulation. Assuming good input data, a lot of qualifying data can be generating from the resulting statistical calculations, such as the minimum and maximum values, the mean, the median, the mode, the standard deviation and the percentiles, see Figs. 1.2 and 1.3. The input parameters have major effect on the method results including the most likely and maximum values for the parameters.

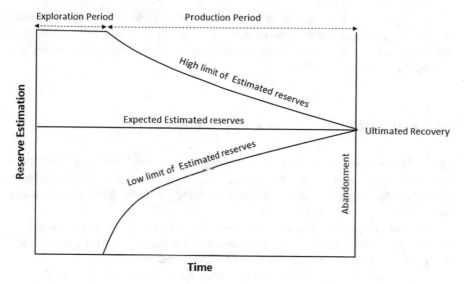

Fig. 1.1 Scale of uncertainty in reserves estimates

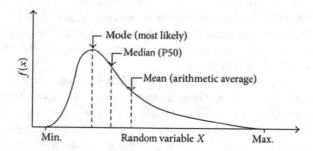

Fig. 1.2 Measures of central tendency

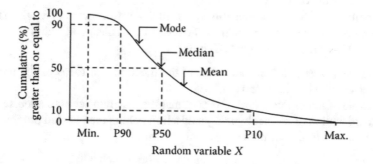

Fig. 1.3 Prospective resources

A comparison of the deterministic and probabilistic methods, however, can indicate quality assurance for reserves estimation. If the two values are very close, then assurance on the calculated reserves is increased. If the two values are too far, the assumptions need to be revised.

1.2   Exploration and Production Licensing

Exploration and production licenses are limited licenses which are granted in licensing rounds according to the Hydrocarbons Act. Exploration and production licenses are granted for a period of time and can be extended for longer time, which in some countries can reach up to 30 years.

The exploration license is a permission giving to a company or a joint venture permitting them to exploration for commercially possible deposits for the extraction of oil and gas from geographical areas at onshore or offshore.

Many countries assign petroleum exploration and development rights in different fiscal terms; either public bids, licensing rounds, or direct negotiation, and most of the countries use a combination of these fiscal terms. Accordingly, for award there are different conditions: some of petroleum owners adopt quite rigid forms with very limited biddable items that affect the sharing of the lease between stockholders and owner of the natural resource; other owners award rights on the basis of work plans; in others, the whole thing is negotiable.

Typically, the exploration licenses can be initially granted for a period of up to 5 years and may be renewed one time, for up to 5 years. A second renewal, for up to 5 years, is only permitted in special circumstances and where it can be demonstrated that there is a possibility of the licensee identifying natural resources during the period of the renewal. No additional renewals are allowed. However, a license may be granted for a shorter time period, mostly if the planned work program will not require 5 years to complete.

An exploration license application must include a program of work, namely: The nature of the program proposed (office-based activities, on-ground exploration activities, sub-surface evaluation activities);

1. As far as practicable, an indication of the location and focus of the proposed exercises with location maps.
2. An explanation of the nature of targets that the program seeks to allocate.
3. An explanation of the geological justification behind the planned program.
4. An expected timing of the exploration program.

The proposed work program must detail the work which will be carry out for each year of the license. The program must state clearly the ground exploration work and office based activities. It is expected that, the investor would commit to target testing in the first 3 years of the license and for drilling to be started by the end of the 3rd year. Besides, the Rules provide that an exploration license submission must include the expected annual expenditure for each year of the license. The expenditure planned in the license application must be consistent with the proposed work program.

During the term of the license, the natural resource owner may request updated information of the work program to be delivered at a specified date. The investor must comply with any such request.

Generally, the exploration activities contain a high level of uncertainty and risk. The geological models were built based on seismic sections and nearby wells data (reservoir rock and fluid data), this will provide just a conceptual geological models. Therefore, no presence of hydrocarbons can be expected till the first exploratory well is drilled.

In order to acquire projects (concession blocks), exploration and operation oil companies have to participate in a bid round or lease sale when blocks will be open for a certain period of time after paying a signature bonus. The offered blocks vary by size depending on where they are located either onshore or offshore, in the same time, the block size are different from country to other country. Figure 1.4 shows an overview of onshore and offshore lease blocks (Sirte Basin, Murzuk Basin, and Kufra Basin, and Ghadames Basin, Libya).

An understanding of the reservoir geology is really important to for development, production, and management process. This principally include several geological key elements which combined together to produce hydrocarbons from the reservoir. Petroleum geology is concerned with the evaluation the key elements in sedimentary basins. These key elements are presented in Table 1.1.

1.3 Field Development Planning Road Map

A Field Development Plan (FDP) delivers the excellent technical solution and roadmap for optimizing the development and production of a field. The development Plan consist of a high number of parameters associated to the geological and physical characteristics of the reservoir, to the operational planning and the economic scenario. The most importance is the elaboration of methodologies that can lead to better hydrocarbon recovery strategies and improve the profitability of the project. The studies of the development plan provide the essential direction and information to evaluate the project is economic or not for all possible development project scenarios, uncertainties and risks in order to select the best optimal development model. Net present value, actually the key element of decision making process.

Generally, determining the most successful way to develop and produce from an oil and gas field with low risk, has always been a challenge in petroleum industry. This is really true during times of price instability. Figure 1.5 shows the road map of the field development process.

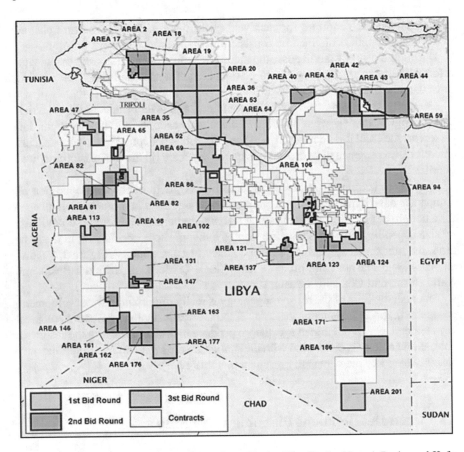

Fig. 1.4 Examples of onshore and offshore lease blocks (Sirte Basin, Murzuk Basin, and Kufra Basin, and Ghadames Basin, Libya. *Source* https://libyanlines.wordpress.com)

Table 1.1 Geological key element

Geological key element	Definition
Reservoir	Is a porous and permeable sedimentary rocks that storing the hydrocarbon reserves
Trap	Underground rock formation that prevents the movement of hydrocarbon and causes it to accumulate in a reservoir
Seal	Is a rock with low permeability that blocks the escape of hydrocarbons from the reservoir rock
Source rock	Rocks from which hydrocarbons have been generated or are capable of being generated

1.4 Exploration and Evaluation

Field development plans give the best technical solutions for field optimization. Figure 1.5 show all activities and processes required to develop a field starting from geology, geophysics, reservoir, well design and construction, completion design, production engineering, surface facilities, infrastructure, environmental impact, and economics and risk assessment. From the discovery of hydrocarbons to the first oil and gas, exploration and production activities are spread over several decades. Several main steps can be distinguished in the life of the field.

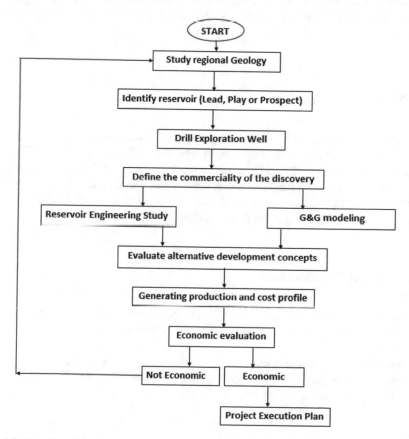

Fig. 1.5 Flowchart showing road map of the field development process

1.4.1   Field Discovery

a. Oil and gas are accumulated underground in reservoir rocks either on shore or offshore,
b. To find underground hydrocarbon accumulation, geoscientists evaluate images of the subsurface geological layers generated by seismic echography. Geophysics and geology (G&G) build static geological model, and classify the potential reservoirs, based on the input data obtained, called Leads/plays/prospects,
c. To defend the commerciality of the discovery, exploration wells need to be drilled to make sure the identified reservoirs contain hydrocarbons or not. However, the discovered hydrocarbon volumes will define the number of exploratory wells needed.

1.4.2   Field Evaluation

After the commerciality of the discovery has been confirmed, reservoir simulation model will be built to estimate the initial hydrocarbon volume in the reservoir, and to simulate the reservoir fluid flow behavior and optimize the field development scenario (type of wells, number of producer and injector wells, and location of wells, capacity of field production, …etc.). Appraisal wells are drilled to improve the reservoir description through getting more data acquisition.

1.4.3   Field Development

a. Defined the number of wells to be drilled to reach target field oil or gas production,
b. Select the optimum recovery methods need to be used to extract the oil and gas from the reservoir,
c. Defined the type and cost of treatment facilities (separator, dehydrator, degasser, water injection facility…etc.) needed either onshore or offshore,
d. Select the treatment systems needed to protect the environment.

1.4.4   Field Production

Normally, the production time period to extract the hydrocarbon from the reservoir varies between 10 and 35 years and could be extended more than 40 years for giant oil and gas fields. The reservoir lifetime is including different continuous phases:

a. A period of production increase,
b. A plateau phase,
c. Secondary phase, injection phases, including water, gas or chemicals to enhance the hydrocarbon recovery,
d. The production depletion period when hydrocarbon production declines with time.

1.4.5 Field Abandonment

Once the field production flow rate is non economical, the reservoir is abandoned. Before abandoning the field, the oil companies need to dismantle all onshore or offshore facilities and plug and abandoned all the wells. The total costs of abandonment operation must be taking into account within the total cost of the economical oil and gas field.

1.5 Economic Evaluation of Oil and Gas Projects

The economic evaluations results are the key elements that are used by directors to make investment decisions. The economic evaluation is the final stage of technical evaluation and financial assessment before any investment decision is made. The economic evaluation process are as follows:

1.5.1 Generating Forecasts of Key Technical and Economic Parameters

The initial step is the collecting of the technical data, which include the following:

a. An annual forecast of the oil and gas production that is expected to be produced by the project. The geologists will provide the amount of estimated oil and gas reserves and the reservoir engineers will generate the annual production forecasting profiles.
b. An annual prediction of the oil and gas prices at which the production is anticipated to be sold. Typically, oil companies approved the predicted oil and gas prices that must be used for economic evaluations process.
c. The expected annual capital expenditures required to explore and develop oil and gas project.
d. The expected annual operating expenses required to maintain the production of the oil and gas.

1.5.2 Modeling of the Fiscal System

Every country has own tax laws and contract conditions that rule the methods by which petroleum companies must pay a portion of their profits to government. Together, these are referred to a country fiscal system. A model must be built that will eventually calculate the annual after tax cash flow that the oil and gas company will obtain from the entire project. Accordingly, from the economic evaluation perspective, there are two key important types of fiscal systems.

1. Royalty/tax regimes. These types of systems are found in countries such as Libya, Norway, United States, Australia, and United Kingdom. The following are the common features:

    a. The oil company "owns" the reserves and therefore takes their profits by selling the crude oil and gas.
    b. The oil company need to pay a royalty to the government, which is normally identified percentage of the revenues. The company must pay an income tax based on the profits of the project (the revenues less the royalties less all costs).

2. Production sharing agreements. This agreement is used in developing markets:

    a. Unlike royalty/tax systems, the oil and gas company does not own all of the hydrocarbon reserves. Reasonably, the government might be considered the original owner of the reserves.
    b. The government assigns a portion of each year's actual oil and gas production, as cost recovery, to the oil company to recover their costs of exploration, development, and operating.
    c. After allocate the cost recovery to oil company, the remaining value of the oil and gas production is the profit of the project. This profit is then shared between the oil and gas company and the government.
    d. There are two main sources of revenues to the oil company, cost recovery and profit share, and two sources of expenses, capital expenditures and the operating costs.

1.5.3 Calculation of Economic Indicators

The final stage in the economic evaluation procedure is to summarize the future cash flow forecasts into many economic indicators that will help to make the investment decision on the project. One of the main key element is the time value of money.

The net present value (NPV) is the best measure of the actual value in monetary terms of any single project. There are several methods used to develop the company discount rate. Besides NPV, there are three other economic indicators:

a. Rate of return (ROR) is derived as the only discount rate that forces the NPV to be exactly equal zero.
b. Payout (or payback) is a significant indicator of how long the project will take to earn its capital expenditures from the profits.
c. A profitability index. This indicator could be done on both an undiscounted or a discounted basis.

There are many other economic indicators might be used, based on its own strengths and weaknesses. When look at together, however, they usually give a very good indication of whether or not the project should proceed.

Chapter 2
Prospective Resources: Plays, Leads, and Prospects

To identify some of the concepts in this book, we need to understand play, lead, and prospect.

2.1 Play

A play is defined as a region thought to be conducive to the hydrocarbon accumulation in a specific geologic formation or interval in the subsurface. Also can be defined as a geographically and stratigraphically delimited area where common geological factors exist in order that petroleum accumulation can occur (Fig. 2.1).

A play initially defined as a perception of a model in the mind of the geologist of how a number of geological features might combine to produce petroleum accumulates at a specific stratigraphic level in the basin. With further investigation the play will be redefined to a trap (petroleum system).

i. **Play types**

- Defined based on the first order closure process.
- Emphasizes the closure as the key element of a play.
- Identified six play families.

ii. **Examples of play types**

- Reef buildup family play (RB) (Fig. 2.1).

  - Main Process: Biological and depositional processes

- Extensional block play family (FB) (Fig. 2.2)

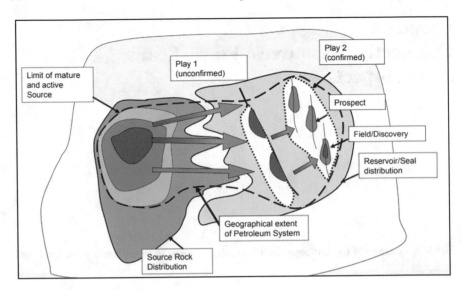

Fig. 2.1 Petroleum system concept

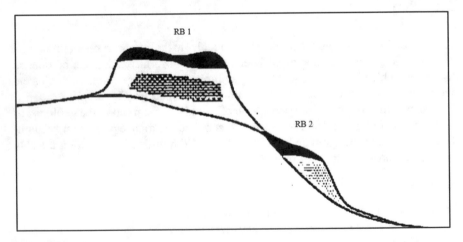

Fig. 2.2 Reef build-up play family

2.2 Lead

Potential accumulation is currently poorly defined and requires more data acquisition and/or evaluation in order to be classified as a prospect (Fig. 2.3).

In general, Lead is any indication or hint of the presence of a trap (a structure which may contain hydrocarbons) in the subsurface which may allow explorationists to explore it further (Fig. 2.4).

Lead is A potential petroleum trap defined by only one seismic line (Fig. 2.5).

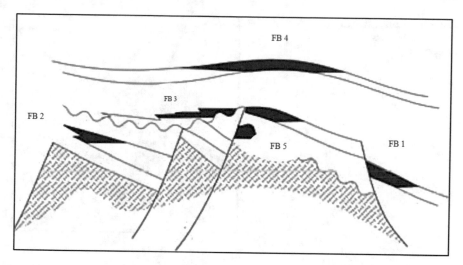

Fig. 2.3 Extensional block play family

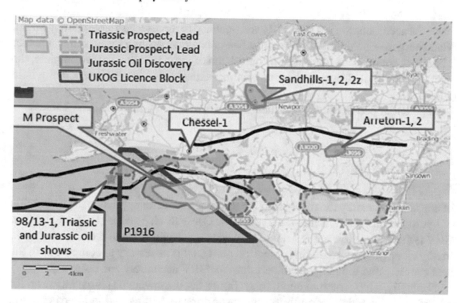

Fig. 2.4 Leads trap. (*Source* http://wig.ht/2g0m)

Fig. 2.5 Map shown one seismic line crossing Lead trap

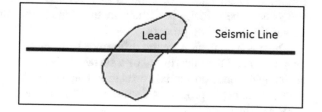

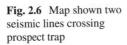

Fig. 2.6 Map shown two
seismic lines crossing
prospect trap

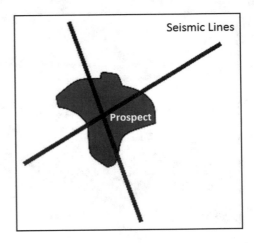

2.3 Prospect

Potential accumulation is sufficiently well-defined to represent a viable drilling target.
Also known as a potential trap that must be evaluated by drilling to determine whether
it contains commercial quantities of petroleum or not. Once drilling is complete, the
term "prospect" is dropped; the site becomes either a dry hole or a producing field
(Fig. 2.6).

A potential petroleum trap defined by a grid of seismic lines (at least 2 lines).

2.4 Rock Volume

In any subsurface trap the most effective technique or factor applied in the estimation
of hydrocarbon volume in the reservoir is the Gross Rock Volume (GRV) technique.
GRV is the volume of rock restricted between a top and bottom of the reservoir rock.
In any oil or gas volumetric investigation process it is necessary to compute both the
best estimate and the range of uncertainty for GRV accurately and correctly.

Normally, a top structure depth map (from a depth conversion) is usually used for
providing a depth prediction to top reservoir prior to drilling and in the calculation
of GRV for use in hydrocarbons initially in place (HIIP). The Time-Depth is very
Important in all phases of exploration, development and production. Figure 2.7,
shows how the velocity is the key factor in the whole study, starting from seismic
survey till the drilling process.

There is a confusion on the difference between the uncertainty at a point (depth
prognosis) and the uncertainty of a surface (GRV estimation). The GRV estimation
critically depends on spatial correlation being reproduced correctly in the mapping,
whereas the depth prognosis does not. This has serious implications for estimating
GRV and its uncertainty.

Fig. 2.7 Flow chart shows the significance of the velocity starting from seismic survey till the drilling process

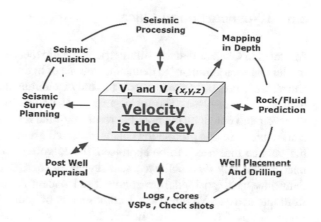

The commonly use of depth surfaces for depth prognosis for well targeting is entirely valid. However, the use of the same surfaces for GRV estimation is not usually valid.

It is very clearly that GRV is the most important parameter in estimating the volume of hydrocarbon in-place. However, compared to other parameters in the HIIP equation GRV is exceptional: there is no tool available that can be directly measured. Therefore, GRV is determined indirectly, using a top structure depth map (Fig. 2.8) and understanding of the trapping mechanism.

Commonly, the workflow for determining a depth map comprises of seismic time interpretation, gridding, time-to-depth conversion and residual mapping. Therefore, uncertainty is existing at all stages including the time interpretation, gridding, estimation of the velocity field and the choice of the method of depth conversion and the intervals to be used.

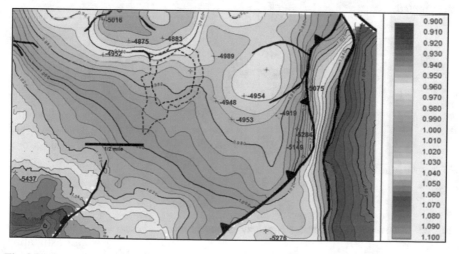

Fig. 2.8 Contoured time-depth map

2.5 Hydrocarbon in Place

As stated in Chap. 1 that the volumetric is a static measurement based on a geologic model that uses geometry to define the volume of hydrocarbons in the reservoir rock. Currently, volumetric estimation is the only technique that is accessible to estimate hydrocarbons-in-place. The main purpose of volumetric estimation is to assess a reservoir and calculate the possible reserves of the reservoir rock. Once drilling has started pressure and production data is collected giving a greater understanding into the volume that needs to be appraised. To do volumetric estimation, geoscientists need to use any data collected such as cores, logs, seismic and other surveys to determine the depositional environment and to identify trap features and to determine fluid interactions. Volumetric is a combination of geological, fluid and the modeled relationships. The flow chart in Fig. 2.9, shows the relations between the different variables.

Amount of hydrocarbon likely to be contained in the prospect (Fig. 2.10).

This is calculated using the volumetric equation (2.1):

$$\text{HIIP} = \frac{\text{GRV NTG Porosity oil saturation}}{\text{FVF}} \qquad (2.1)$$

where,

HIIP = Hydrocarbon Intimal in Place.

GRV = Gross Rock volume—amount of rock in the trap above the hydrocarbon water contact.

NTG = Net/Gross ratio—proportion of the GRV formed by the reservoir rock (range is 0 –1).

Porosity = percentage of the net reservoir rock occupied by pores (typically 5–35%).

Hydrocarbon saturation = some of the pore space is filled with water—this must be discounted.

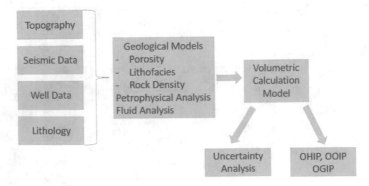

Fig. 2.9 An overview of the interactions between the different variables

Fig. 2.10 Map shown the areas of hydrocarbon likely to be contained

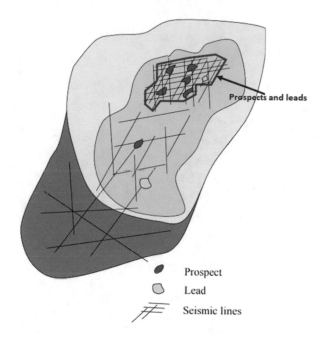

Prospects and leads

Prospect

Lead

Seismic lines

FVF = formation volume factor—oil shrinks and gas expands when brought to the surface. The FVF converts volumes at reservoir conditions (high pressure and high temperature) to storage and sale conditions.

Typically, volumetric estimates provide a static quantity of the initial hydrocarbons in place. This means that the precision of these quantities are greatly dependent on the amount of data the geoscientist has obtainable to them. In the early phases of development, the data of a potential reservoir is limited and the volumetric estimate is based on the accuracy of the parameters such as GRV, reservoir geometry and trapping, pore volume, permeability, and fluid contacts.

Once the phases of exploration continue and drilling operation begins, the accuracy of the volumetric estimation will increase. Besides, the data obtained during drilling gives a better representation of the reservoir properties which might increasing the accuracy of the volumetric estimation.

Chapter 3
Understanding Exploration, Appraisal and Production Stages

Exploration is a costly and risky operation as the expenditure associated are generally valued at millions of dollars and every two out of three wells, on average, contain no shows of hydrocarbons. Consequently, the exploration companies need to drill more wells in one area to find an oil or gas discovery, and this might take more years. There are many process need to review before to decide to drill or not such as basin-scale conventional play assessment:

1. Identify areas of a basin where there are: source rocks, reservoirs and traps (petroleum system).
2. Identify prospects in those areas.
3. Rank the prospects by risk.
4. Drill the best one, then re-evaluate the others.

During exploration drilling, significant information and reservoir rock and fluid samples are collected which are encountered by the well in order to know if there any hydrocarbons are discovered at drilled location.

Sometimes, some exploration companies they find nothing after long exploration operations. Figure 3.1, shown the dry holes of oil and natural gas wells drilled from 1985 to 2000 in Gulf of Mexico. Clearly, the number of dry holes declined as technology improves.

Normally, after the prospect has been carefully chosen and is ready to be drilled, an initial exploratory well is created. This well will discover the accumulation and prove if there are any hydrocarbons existing in the reservoir as per the assumptions made.

The exploration stage includes the work prepared by geoscientists to interpret seismic sections, build top and base structure maps, and define the location for an exploratory well and subsequent appraisal wells.

© The Author(s), under exclusive license to Springer Nature Switzerland AG 2020
T. A. O. Ganat, *Technical Guidance for Petroleum Exploration and Production Plans*,
SpringerBriefs in Applied Sciences and Technology,
https://doi.org/10.1007/978-3-030-45250-6_3

Fig. 3.1 Shows the dry holes as a percentage of oil and natural gas wells drilled from 1985–2000 in Gulf of Mexico (https://www.energyandcapital.com/articles/drilling-the-bakken-formation/73660)

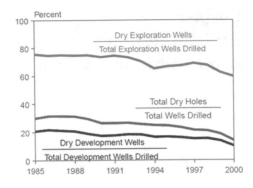

3.1   Exploratory Wells

Exploratory wells are drilled in regions that have not been previously discovered to contain hydrocarbons (yet undiscovered pool). It includes a relatively high level of risk. They could also be drilled in areas near drilled wells to find another reservoir contains of oil or natural gas (Fig. 3.2).

Exploratory wells are drilled in places that have undergone seismic testing to define the depth and thickness of possible sources of natural gas. After drilled wells, engineers study the various rock layers to conclude which layers are possibly sources of natural gas. Normally, the exploratory wells are drilled vertically but it can be drilled as horizontal wells only if the well is expected to be productive (Fig. 3.3).

Typically, a prospect will have one or two exploratory wells, depending on its extent and degree of reservoir compartmentalization. Exploratory wells are planned to test only the specific given area and not the whole prospect area. Exploratory wells might be categorized as (1) wildcat, drilled in not proven area; (2) field extension; or (3) deep test, drilled within a field area but to not proven deeper layers.

As the well proved the availability of hydrocarbons at the chosen area, other wells called appraisal wells need to be drilled to define the extension of the reservoir area and the fluid contacts.

Fig. 3.2 Simple structure drilled with one exploratory well

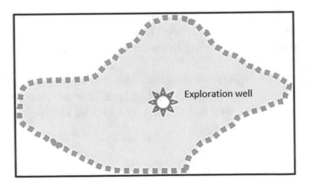

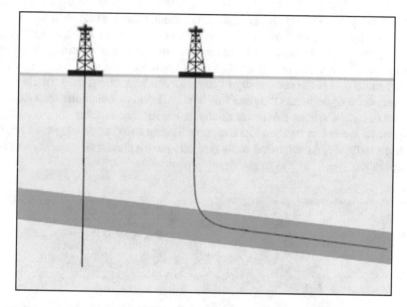

Fig. 3.3 Diagram shows vertical and horizontal well

3.2 Appraisal Wells

Appraisal Wells are a vertical or deviated wells that are drilled into a discovered hydrocarbon accumulation to further understand the extent of the hydrocarbon accumulation before commercial production of hydrocarbons from a well can commence (Fig. 3.4). The drilling of these wells is very significant part of the exploration and production operations of any exploration and production company. In the sequence of the development program, appraisal drilling is executed before starting the commercial production.

Appraisal wells are drilled in order to prove that the discoveries are commercial and give an economic appreciation to the exploration and production companies.

Fig. 3.4 Simple structure prospect not compartmentalized with one appraisal well

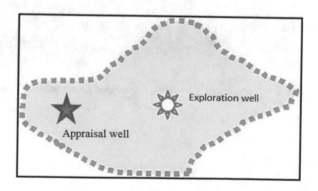

Wells can be drilled either before or after an accumulation has been developed. After drilled the appraisal many information can be obtained such as reservoir size and volume, possible production rate from the reservoir, flow and volume of fluids, etc.

Normally, it's difficult to defined how many appraisal well need to be drilled. The reason behind that is the complexity of the reservoir structure (Fig. 3.5). Therefore, engineers need to delineate the reservoir boundaries to calculate the total expected volumes of hydrocarbons that are restricted in the reservoir rocks.

Appraisal wells can be abandoned after drilling or saved as development production wells in future. An appraisal wells typically have a chance of success more than an exploration wells but less than a development wells.

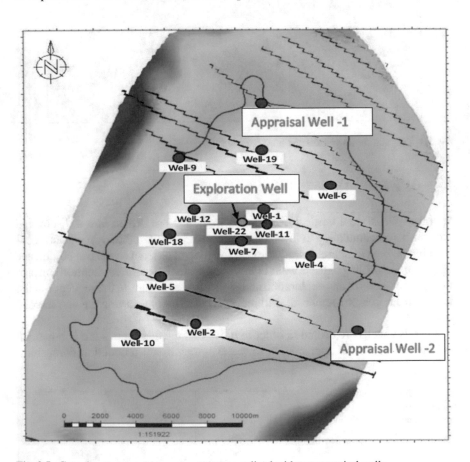

Fig. 3.5 Complex structure prospect compartmentalized with two appraisal wells

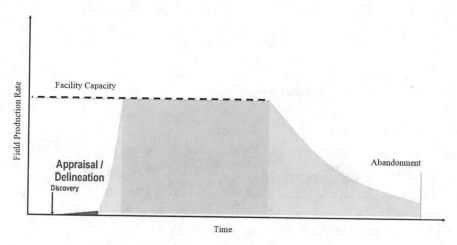

Fig. 3.6 Appraisal and delineation Stage of the oil and gas reservoir or field

3.3 Appraisal and Delineation Phase

Appraisal and delineation wells are wells drilled to obtain more information to improve the understanding of the reservoir rock and fluid properties (Fig. 3.6). At this stage, a rig schedule is normally developed with input from most asset team members, such as drilling engineers, the development geologists, and the reservoir engineers. The rig schedule orders the wells to be drilled based on the goals and the logistics of each well.

Throughout the appraisal and delineation period, both drilling engineers and development geologists will work together to generate the drilling proposals. In this stage, the rig schedule is full with many drilling activity; however, plugging and abandoning (P&A) and temporary well suspension activities are also involved.

If many drilling rigs are existing for the project, then the different appraisal and delineation wells in the rig schedule are place in order and allocated. Also, if a single rig or multiple rigs are servicing many projects, then gaps in the rig schedule are allocated to a project to permit for the timely interpretation of data from a well. Therefore, the next proposed wells to be drilled can get benefit from the data obtained.

3.4 Scheduling for Exploration and Appraisal Phases

The exploration and appraisal stages of a certain project will differ of course dependent on the volume, number of wells need to be drilled, and the time required to drill all the proposed wells. Normally, production and operations companies will drill an exploratory well first than drill the appraisal wells over a period of 2–4 years. Commonly drilling one or a maximum of three wells per year, mainly for Deepwater

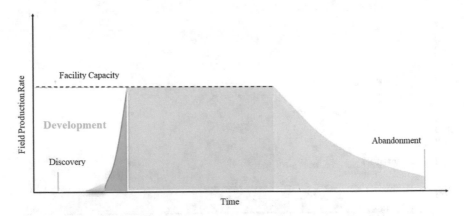

Fig. 3.7 Development stage of the oil and gas reservoir or Field

offshore wells, within the depth ranging from 13,000 to 30,000 ft below the seabed. The deeper offshore well, the more expensive it will be and the more time required to drill the well.

3.5 Development Stage

In the development stage, wells are drilled with the main objective of hydrocarbon production. This is the stage in the field where the development plan is executed. There is a strong economic motivation for developing the field.

During the development stage of oil and gas field (Fig. 3.7), the rig schedule is practically fully committed to drilling the wells considered in the original field development plan. During the period of the development drilling, the learning curve becomes steeper and in the same time the drilling times usually become smaller as the drilling engineers and rig crews used their experience gained from previous drilled wells.

3.6 Petroleum Plateau Stage

Plateau stage means the period that is specified in the approved development plan in respect of an oil field or a gas field. During the Plateau Stages of the field life-cycle, multiple reservoir management activities occur. In the plateau period most of the planned development wells were drilled and tied–in to the production line (Fig. 3.8).

During this phase, the rigs are often used for workover activities. These workovers require the use of the rig to perform well remediation activities that cannot be performed using a slick-line, Wireline, or coiled tubing. Workovers operations includes

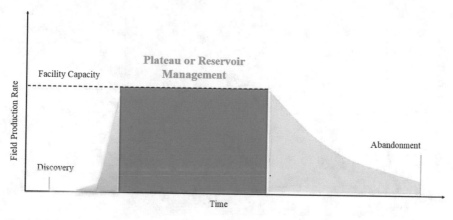

Fig. 3.8 Plateau period of the oil and gas reservoir or Field

well completion, zonal isolation, pump installation, gas lift pumps installation, fishing jobs, acid job, fracturing…. etc.

3.7 Decline Phase

During this stage, the production performance of the reservoir or the field start depleting (Fig. 3.9), due to reservoir pressure depletion or any other production strategy used such as an increased water cut will cause a decline of the oil production flow even with high reservoir pressure (Fig. 3.10).

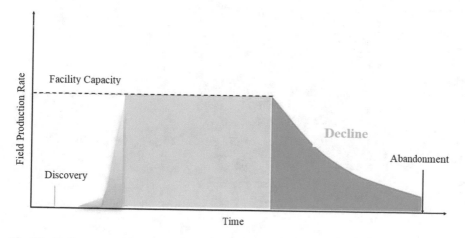

Fig. 3.9 Decline stage of the oil and gas reservoir or Field

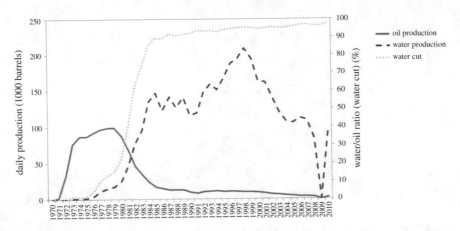

Fig. 3.10 shows increased water cut will cause a drop of the oil production flow at high reservoir pressure

Later in the lifetime of the reservoir or field, the drilling of additional new wells could be required after all phases of the original field development plan have been finished and the reservoir management activities focus on arresting the reservoir decline. For instance, to arrest the declining oil or gas production flow rates, Infill drilling may be applied. Infill drilling is approach where the original wells drainage areas are decreased by drilling new wells between the existing development wells.

In oil reservoirs, pattern rearrangement might be used as a secondary recovery processes, such as gas or water injection to adjust the producer-injector patterns to produce any oil volume not recovered during the primary recovery method.

Chapter 4
Estimation of Reservoir Rock and Fluid Properties

Reservoir rock and fluid properties are the vital parameters that control the estimation of the original oil and gas in place and the recoverable oil and gas.

During the evaluation process of the hydrocarbons in place, it's very important to understand the distribution of the reservoir rock and fluid parameters along the extend reservoir. There are several approaches to identify the reservoir characterization more clearly which can help to analyze more accurately all the required reservoir rock and fluid parameters to calculate the hydrocarbons in place by using the formula in Eq. 2.1 in Chap. 2.

4.1 Porosity/Depth Trends in Reservoirs

Exploration in very deep reservoirs needs better predictions of porosity/depth interactions. These must be based on the temperature and pressure associated to the burial history, and on the estimations of the primary reservoir mineralogy and texture. Typically, the porosity reduction at depths down to 2.5–3.0 km, is due to mechanical compaction [23] and the primary sorting and the clay content are significant factors [26]. At larger depths, quartz cementation becomes extremely important [3, 9, 10, 22, 23] and porosity dropping is almost related to the processes governing cementation of the quartz.

Porosity trends with depth in reservoir rock have been proposed in many published papers [2, 4, 5, 13, 16, 19, 25, 26]. Also, there are many other parameters affecting the gradient of porosity/depth trends have been clearly presented. Age and timing of petroleum emplacement have been addressed by many researchers.

Normally, the porosity is acquired from the wells already drilled in the exploration areas. Commonly, porosity/depth trends provide a good basis of understanding the porosity estimation ranges at different depths in a specific reservoir.

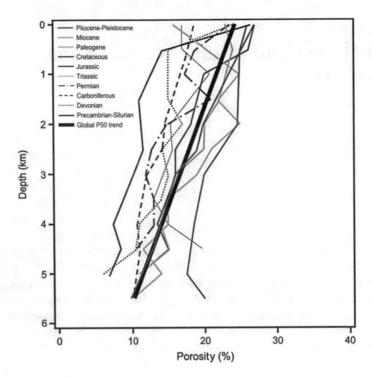

Fig. 4.1 Porosity/depth trend for siliciclastic reservoirs

Figures 4.1 and 4.2 show two porosity/depth trends for siliciclastic reservoirs and carbonate reservoirs. These plots can be utilized to predict the ranges of the porosity for a play or prospect located at a specific depth.

4.2   Porosity-Permeability Trends

Porosity and permeability are measured in the laboratory using routinely method on small core samples cut from the main core obtained from the reservoir rock. Once perm-poro data from a given reservoir rock are plotted, the resultant trend normally is used to understand permeability distribution from porosity, which is more certainly estimated from well logs than permeability. However, data from different reservoir rocks form trends at different points on a log (perm-poro) plot.

By comparing porosity/permeability trends from several wells it is likely to estimate the reservoir permeability distribution between wells. Permeabilities values predicted from porosity/permeability trends should be considered as qualified but not essentially accurate predicts. Understanding deviations from perm-poro trends are significant to reduce the risk of predicting inaccurate permeabilities.

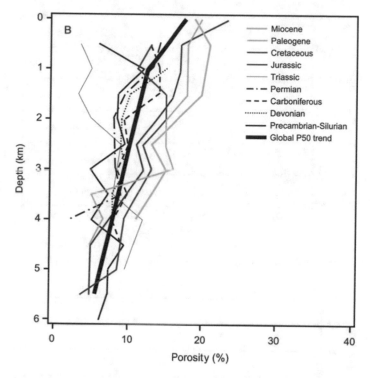

Fig. 4.2 Porosity/depth trend for carbonate reservoirs

Following the same porosity/depth trends methodology, it is also possible to create a range of permeability from porosity assuming there is a suitable trend between the permeability and porosity.

Figure 4.3 is porosity/permeability plot for the sandstone reservoir with the trend line of the sandstones plotted for 375 samples for comparison. The plot shows the trend rises with greater porosity providing higher permeability. The permeability is not needed in the hydrocarbons in place calculations but it will have an effect on the reservoir recovery efficiencies and well flow rates.

4.3   Water Saturation/Permeability Trends

The obtained permeability and water saturation data from the reservoir can provide the trend and the distribution of these parameters in the reservoir. Typically, the higher the initial water saturation, the lower the permeability, as more water will filled in the small pores or network of pores.

Figure 4.4 shown the trend between initial water saturation and permeability. The figure shows the scattered data are very consistency and it's a good agreement with a

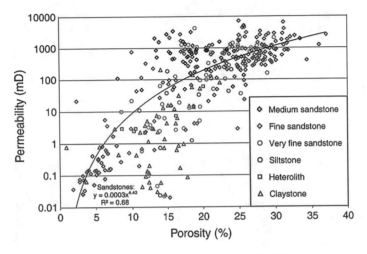

Fig. 4.3   Porosity/permeability plot for the sandstone reservoir with 375 samples

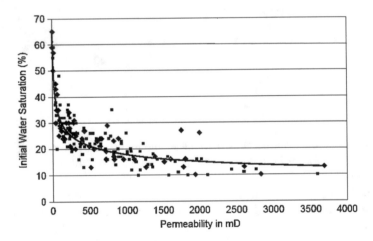

Fig. 4.4   Trend showing a good correlation between the permeability and initial water saturations

fitted power function. This gives us more confidence to use it in the lack of reservoir data or even to validate data from other sources such as core data or laboratory core measurements.

4.4   Recovery Efficiency Versus Depth/Play Trends

Recovery efficiency is a measure of the volume of resource recovered relative to the volume of resource originally in place. An assessment of predictable recovery efficiency can be acquired by known other factors that has influence on the recovery of a reservoir fluid. Two main factors are very significant, volumetric sweep efficiency and displacement efficiency. Where, the displacement efficiency describes the efficiency of recovering mobile hydrocarbon and volumetric sweep efficiency describes the efficiency of mobile hydrocarbon in terms of areal sweep efficiency and vertical sweep efficiency. The efficiency of primary recovery mechanisms will differ generally from reservoir to reservoir, but the efficiencies are usually highest with water drive, intermediate with gas cap drive, and with solution gas drive.

The recovery efficiency is a key parameter which the exploratory team need to estimate at early time of the project to get an initial expected range for hydrocarbon recoveries.

Figure 4.5 shows recovery factors versus depth for oil reservoirs and Fig. 4.6 for gas reservoirs, for different prospects in the deepwater reservoirs. The figure shows that the recovery factors for oil reservoirs, are varying from 10% to 40%. The recovery factor for Miocene reservoirs and for lower Miocene reservoirs show a

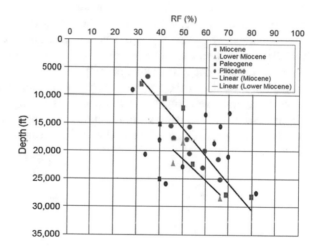

Fig. 4.5 Recovery factors (RFs) for deepwater oil reservoirs

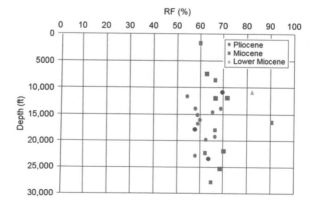

Fig. 4.6 Recovery factors (RFs) for deepwater gas reservoirs

good trend with depth. The Pliocene reservoirs are the only ones that not following a specific trend. More work should be done to set the Pliocene sands according to their reservoir permeability features.

Where at high permeability the recovery factor varying from 30% to 40%, while at low permeability its varying from 10% to 25%.

Figure 4.5 shows the recovery factor for the gas reservoirs which is varying from 50% to 90% with an average around 60% to 70%. This low recovery might be due to the existence of gas condensate. No specific trend can be observed. However, the ranges seen in the figure can be used for initial screening of gas recoveries for the prospects and depths shown.

4.5   Formation Volume Factors

This factors, are used to convert the flow rate of oil or gas at stock tank conditions to reservoir conditions. These factors can be obtained by using empirical correlations such as Standing correlation and Vasquez and Beggs correlation or from laboratory PVT measurements. The value of the oil formation volume factor is generally between 1 and 2 Rbbl/stbbl (R m^3/st m^3).

It is preferred that the formation volume factors measurements are performed at initial average reservoir pressure conditions during the time of the well test. If empirical calculations were used to obtain the formation volume factors than, keep in mind that the solution gas oil ratio has a significant effect on the formation volume factor value.

Chapter 5
Petroleum Reservoir Analogues

Obtaining geophysical data is a time consuming and costly activity. Also, depending on the exploration stage, there is few data available and it is not even possible to obtain new data (for the duration of the bidding phase). One of the general strategies in the industry is to use existing known information from similar reservoirs (analogues) to better determine the missing data, to get more information about the target area. Therefore, the technical team, geologist and reservoir engineers must evaluate all accessible data types such as seismic, well logs, well testing, production data, cores and analogues data available, to make the best technical and business decisions in exploration and production operations process. Therefore, a lack of resources or time limitations mean this requirement is not fully carry out; then the finally decision could lead to drill dry holes, and ineffective field developments plan. The proper way to decrease uncertainty of the obtained data, is to use the data from analogue oil and gas fields and reservoirs as references and benchmarking purposes.

5.1 Using Nearby Analogues

The producing oil and gas fields provide good analogues, permitting for a direct assessment between the new exploration opportunity and the existing producing fields which have been attained from analogous exploration plays and prospects (Fig. 5.1). Typically, the selected analogue is often referenced because it looks reasonably similar and its located in the same basin or geographic region. The use of only one analogue carries high risk and it is not strong enough benchmark to describe any new exploration opportunity.

The main critical part of any analogue databank is consistency for all the technical parameters such as basin type, environment depositional system of the reservoir, structural setting, reservoir properties and thickness, the reservoir lithology, trapping mechanism, seal lithology and shape…etc.

© The Author(s), under exclusive license to Springer Nature Switzerland AG 2020
T. A. O. Ganat, *Technical Guidance for Petroleum Exploration and Production Plans*,
SpringerBriefs in Applied Sciences and Technology,
https://doi.org/10.1007/978-3-030-45250-6_5

Fig. 5.1 Evaluation process using analogue database

To date, the main weakness is the inability to directly describe the new exploration opportunity (play, prospect or field development) against commercially available analogue records.

5.2 Using Global Analogues

Typically, global analogues defined as a large number of exploration blocks from different regions that had huge number of wells and seismic data, which aligned based on the regional producing trend. By using search parameters based on the tectonic structural setting of the blocks, new discovery opportunity can be characterised based on many similarly producing reservoirs within the region (true analogue data).

Engineers have the choice of picking from a combination of many available technical parameters for a tabulated full play characterisation, which permits screening economics at early stages of the evaluation cycle to estimate least possible economic field size. Also, it is able to highlight any potential production difficulties that can add to the complication of any future field development plan which might increase the project costs.

5.3 Improved Hydrocarbon Recovery

Reservoir engineers can find all fields that have a similar geological and engineering parameters and developed over secondary and tertiary recovery production phases. The performance of production development cases can be easily retrieved and studied. So, the engineers can use these data and compare the new developed production field performance versus existing analogues fields and define the most effective production improvement approaches and determine the future performance for the new fields.

5.4   Analogue Data in the Reservoir Modelling Workflow

Analogue data are significant in the geological and reservoir modelling process for understanding the geometry, dimensions and the distribution of the sedimentary heterogeneities and sandstone bodies that contain and control hydrocarbon flow [1, 7, 12, 14, 15, 31]. The technique in which analogues are used depending on the forming workflow, the phase within that workflow and the purpose of the model. Figure 5.2 is a general sketch of a typically modelling workflow and the kinds of data that can be utilized at any stage.

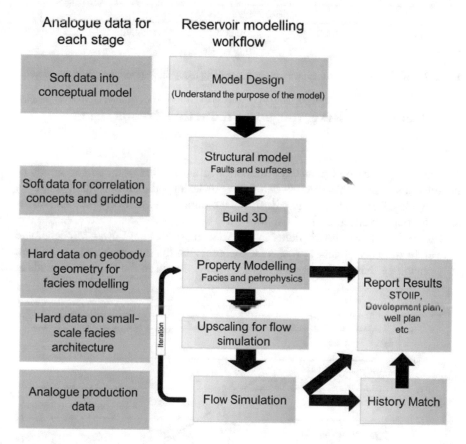

Fig. 5.2 Typical reservoir modelling workflow with different sources of analogue data

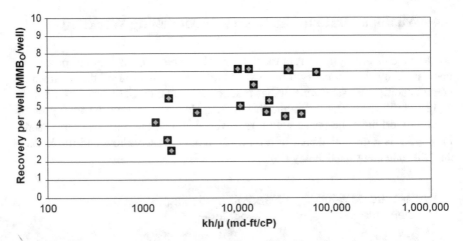

Fig. 5.3 An example of recovery per well trend deriving using analogs and grouping parameters

5.5 Generating Reservoir and Production Trends Using Analogy

By grouping reservoirs or wells parameters in a way that you can see trends or correlations, where these trends can use to make or validate the initial proposed assumptions. For instance, Fig. 5.3, shows the trend of oil recovery per well versus kh/μ for many wells at a certain area.

Figure 5.4, shows the oil fields in the Miocene, Pliocene, Lower Miocene, and Lower Tertiary (Paleogene) prospects. By observing the data scattered on the plot, it is obviously that for the lowest recovery factor found in the Paleogene sands at

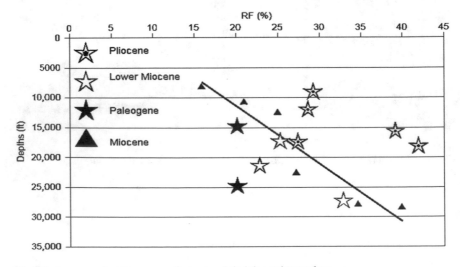

Fig. 5.4 An example of a recovery factor trend deriving using analogs

different depths and the highest recovery factor was in the Pliocene sands. So, such trends can be used to determine any recovery factors for any discovery opportunity which is located at the same depositional age. Besides other trends examples such as recovery versus oil quality or versus a group of fluid and rock properties can be created such as Figs. 5.5 and 5.6.

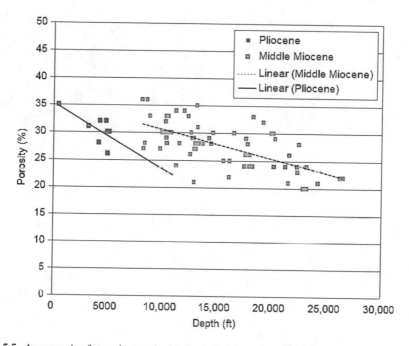

Fig. 5.5 An example of porosity trend with depth deriving using analogs

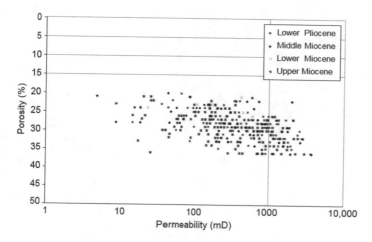

Fig. 5.6 An example of porosity versus permeability trend deriving using analogs

Chapter 6
Wells and Production Functions

After identifying potentially feasible fields and at the beginning of an oil field development plan, once the oil field dimensions have been allocated, engineers determine the number of wells needed to meet production requirements and the method of extraction of the liquid hydrocarbons. Assuming homogeneous rock properties and ideal geological conditions (uniform and continues reservoir), the ultimate primary recovery is independent of well spacing [6]. Muskat [18] envisaged the well spacing concepts based on the physical ultimate recovery and the economic ultimate recovery. From the physical viewpoint there is a minimum number of wells needed to reach maximum hydrocarbon extraction. If the number of wells increased beyond this number, then the ultimate primary hydrocarbon extraction will not increase. From the economic ultimate recovery viewpoint, giving no time limit for the life of the reservoir development project. Logically, at one extreme a few wells can extract the whole hydrocarbon in the reservoir, and at the other extreme, an excessive high number of wells can efficiently extract the hydrocarbon from the reservoir more quickly, but at a high cost. In both cases, the economic return of the project would be negatively affected. Between these two excesses there would be an ideal number of wells that would create maximum economic return. This model applies similarly for vertical wells and horizontal wells.

In fact, most of the reservoirs are not very homogeneous. Typically, the well spacing has been estimated graphically from a plot of economic return verses well spacing as recommended by Muskat [18]. A method to determine the optimum well spacing direct without a plot was proposed by Tokunaga and Hise [28].

6.1 Optimum Well Spacing

Drilling extra wells will increase the productivity for every well, as tighter well spacing speed up the recovery of reserves, which is preferred by project economics.

© The Author(s), under exclusive license to Springer Nature Switzerland AG 2020   41
T. A. O. Ganat, *Technical Guidance for Petroleum Exploration and Production Plans*,
SpringerBriefs in Applied Sciences and Technology,
https://doi.org/10.1007/978-3-030-45250-6_6

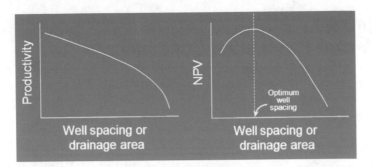

Fig. 6.1 Well spacing and drainage area considerations

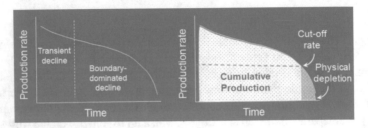

Fig. 6.2 Production decline and cumulative production for a vertical well

Though, it will also increase the drilling expenditures. For this purpose, Net Present Value (NPV) will reach the optimum well spacing (Fig. 6.1).

Most of the wells reach an economic rate during boundary dominated flow. At the cut-off rate, the well spacing area is very close to physical depletion time (Fig. 6.2). Accordingly, if the productivity or ultimate recovery is determined via the economic cut-off rate, it will be almost equal to that determined by using physical depletion. Consequently, physical depletion can be used as a foundation for recovery factors and project economics.

Alike to conventional wells, well spacing is key factor affecting the economics of unconventional wells. It is typically proposed that the physical drainage area of fractured horizontal wells is a function of the well spacing and the bulk of the stimulated reservoir volume (SRV). For instance, in Fig. 6.3, the effectiveness of the depletion of the reservoir between fractures (stimulated reservoir volume) depends on the efficacy of the stimulation, which is a function of the number of fractures along the well section. This means, the number of fractures affects the recovery from the reservoir volume between fractures. Therefore, to increase the recovery efficacy of the volume between the two stimulated reservoir volumes in Fig. 6.3, can be performed by increasing the lengths of the hydraulic fractures, $2X_F$, by decreasing the distance between two horizontal wells, d_W.

Fig. 6.3 Diagram shows two fractured horizontal wells delimited by stimulated reservoir volumes in an unconventional reservoir

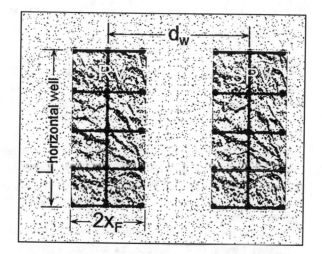

Fig. 6.4 Shows the production decline and cut-off rate for a fractured horizontal well in an unconventional reservoir

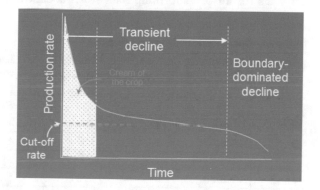

There are so many fractured horizontal wells in unconventional reservoirs, run to an economic rate during transient flow (Fig. 6.4). So, the method to directly relate recover to economic rates is of practical interest.

The actual complexity in optimizing the fracture and well spacing in unconventional reservoirs is the project NPV is ruled by the high productivity at early production times, which rapidly drops to much lower but constant level for the rest of the production time (Fig. 6.4).

Normally, fractured horizontal wells in tight shale-gas prospects, the flow from the outer well spacing (beyond SRV) is insignificant because of the low matrix permeability. In such cases, the well spacing of the horizontal well is the boundaries of the wells' SRVs contact each other.

6.2 Finding the Optimum Number of Wells-Multi-lateral Depletion

Varying the well spacing of lateral wells, the following example was designed for five drilling scenarios with fixed drainage area (Table 6.1). The optimization results propose that the maximum production goes to 1,000 feet well spacing between the laterals (Fig. 6.5). The second best productive drilling scenario might be the 4 wells scenario with 1,250 feet well spacing.

Generally, we judge the success of a project according to the NPV. Accordingly, Fig. 6.6 compares the best NPV for the drilling scenarios. While the 5 wells scenario produces the maximum NPV, and the 4 wells scenario produces the second optimum NPV. Also, the 4 wells scenario can be an adequate choice, mainly if the capital budget for drilling and completion is limited.

The short term production in development plans could be significant in returning the initial capital costs. Figure 6.7 compares the recovery factor after 5, 30, and 50 years for all of the drilling scenarios. As shown in the figure, the five wells

Table 6.1 Optimized completion parameters

Scenario	Well spacing (ft)	No of wells	Fracture spacing (ft)	Fracture conductivity (md-ft)	Fracture half length (ft)
1	500	8	350	50	250
2	750	6	350	46	350
3	1000	5	250	49	400
4	1250	4	250	51	400
5	1500	3	250	50	400

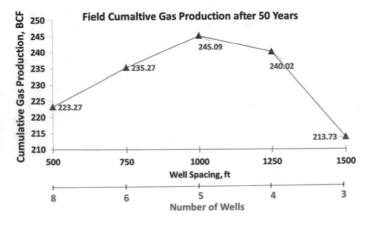

Fig. 6.5 Field cumulative gas production versus Well spacing for five drilling scenarios

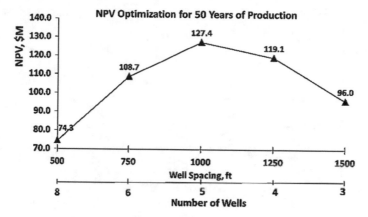

Fig. 6.6 Shows Net Present Value versus well spacing for five drilling scenarios

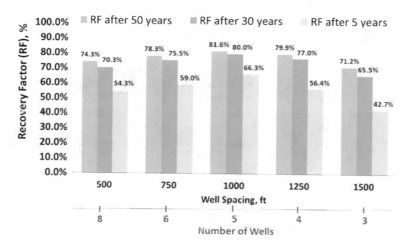

Fig. 6.7 Field recovery factor for all the drilling scenarios at different time periods

scenario at 1,000 feet well spacing has the maximum recovery factor in all three time periods. Also, considering the 5 year recovery factor, the six wells project built at 750 feet well spacing might increase the recovery factor than a four well scenario at 1,250 feet well spacing. But the limitation of the capital budget must be considered. The main observation concluded that all the scenarios, produce more than a 50% recovery factor after just 5 years of production period, except the three wells (1,500 feet) scenario. This is mostly due to the production of wells is limited to the SRV.

Figure 6.8 compares the CAPEX and OPEX costs of these scenarios against the NPV values. As it is anticipated, a greater number of wells will make a higher capital cost. Though, completion of the projects usually needs a considerable share of CAPEX cost. In addition, a greater production will lead to increase the OPEX cost.

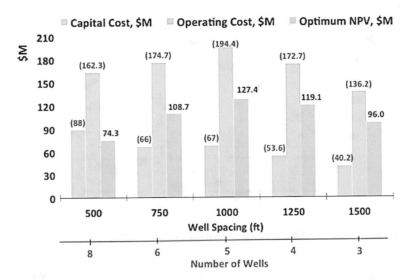

Fig. 6.8 Comparison of CAPEX and OPEX costs versus NPVs for all drilling scenarios

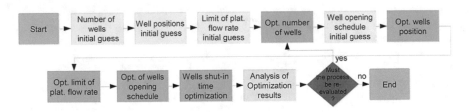

Fig. 6.9 Workflow used to optimize the number of wells

6.3   Optimization Number of Wells

The optimization workflow starts with the guess of number and location of the wells and the limits of the platform production facilities (Fig. 6.9). These process are required to initialize the optimization workflow. By obtained the required data such as field recovery factor and volume of hydrocarbon in place, number of producers and the total hydrocarbon production rate can be expected along with the estimated number of the injectors. Also, based on the quality maps of fluid saturation distribution, locations of the producers and injectors can be allocated.

Chapter 7
Upstream Facilities and Subsea Production Systems

7.1 Facilities Design and Capital Costs Considerations

This chapter presents upstream production facilities and subsea engineering design of oil-handling systems and along with the capital expenditure costs. Several operations activities will be covered including floating production facilities, tiebacks flow lines, types of hosts, export pipelines, and manifolds. Besides, will know how to choice and consider the optimum facilities and evacuation systems.

Normally, the production systems and hydrocarbon storage units in offshore deepwater or ultra-deepwater, will use floating production system, floating production storage offloading vessel (FPSO), semisubmersible platforms such as tension leg platform (TLP), compliant platform, gravity concrete platform, spar platform...etc (Fig. 7.1) as a standalone host. These types of platforms will give more options to select the optimum platform in remote areas where no subsea tieback is possible. Typically, the subsea tieback is an option which will give a solution when the wells and manifolds are extant more than 25 miles away from the main production treatment facilities (Figs. 7.2 and 7.3).

The capital expenditure (CAPEX) involved for the production treatment facilities will be much higher than that for a subsea tieback, where no many subsea facilities are required. Normally, tieback will involve a tariff to be paid if another operator's facilities are used to transport the production.

Table 7.1 and Fig. 7.4 shows CAPEX expenditures for different facilities as an examples of deepwater fields. Table 7.1 shows the variations of the total costs from 600 to 1800 MMUS$, which is highly depending on the total production capacity.

Figures 7.4 and 7.5 shows the cost values for facilities' CAPEX which can be used as guidelines for the future projects that have the peak rates are located within these ranges.

Similarly, we need to estimate subsea and pipelines costs, which are depending on the number of wells (manifolds) and length and diameter of pipelines which will be covered in Chap. 8.

© The Author(s), under exclusive license to Springer Nature Switzerland AG 2020
T. A. O. Ganat, *Technical Guidance for Petroleum Exploration and Production Plans*,
SpringerBriefs in Applied Sciences and Technology,
https://doi.org/10.1007/978-3-030-45250-6_7

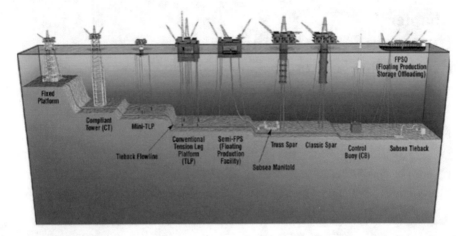

Fig. 7.1 Shows types of fixed and floating offshore platforms (*Source* DNV GL Oil and Gas)

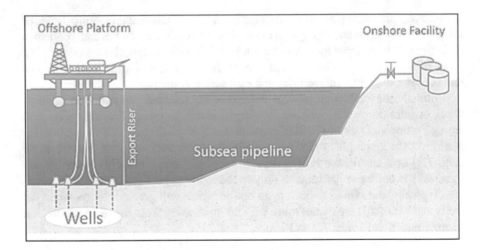

Fig. 7.2 Schematic of subsea pipeline

Figure 7.6 shows, an example of the present network of pipelines in the offshore area. As seen, most of the deepwater areas are not extended by this infrastructure. So, any new deepwater projects located away of these network, required to build new production treatment facilities and a pipeline construction.

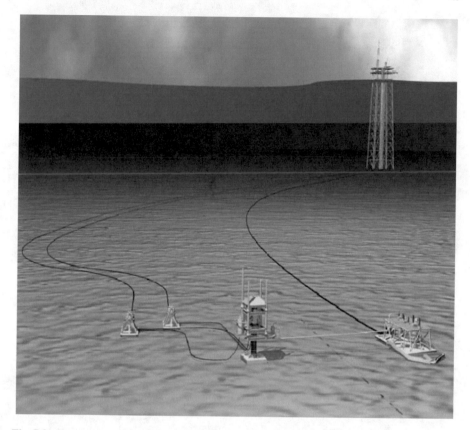

Fig. 7.3 Shows subsea tieback (*Source* Shell Exploration and Production-GOM Operations)

Table 7.1 Facility CAPEX costs versus facility maximum rates for deepwater fields

Field (Facility)	Maximum production rate (Mbopd)	CAPEX (MMUS$)
Field-A (TLP)	70.00	1000.00
Field-B	45.00	900.00
Field-C (SPAR)	100.00	600.00
Field-D	120.00	1500.00
Field-E	135.00	1800.00
Field-F (TLP)	150.00	942.50
Field-G	75.00	700.00
Field-H	250.00	5000.00

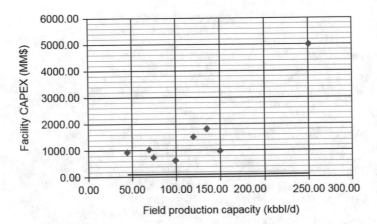

Fig. 7.4 Facility CAPEX versus facility maximum rates for deepwater fields

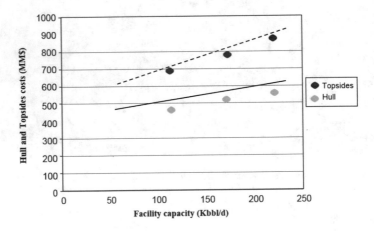

Fig. 7.5 Facility costs subdivided into hull costs and topsides costs

7.2   Wet Tree Versus Dry Tree

For the dry tree system, trees are placed close to the production platform, while wet trees can be located anywhere in a field area (Fig. 7.7). Worldwide, more than 70% of the wells in deepwater are used wet tree systems.

Wet tree systems: Typically, subsea cluster wells, gathers the production in the well-organized and economical way from close subsea wells, or from a remote subsea tie-back to the existing production facilities (Fig. 7.8).

The wet tree system benefits:

- Tree and well access at the seabed isolated from people
- Full range of hull types can be used

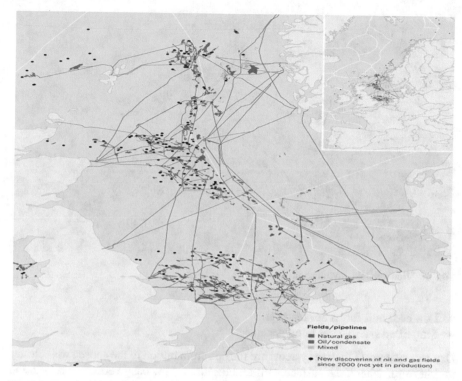

Fig. 7.6 Extant offshore infrastructure of pipelines in the deepwater area (*Source* https://qsr2010. ospar.org/en/ch07_01.html)

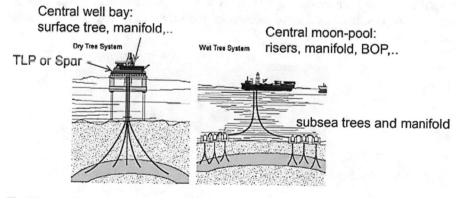

Fig. 7.7 Dry Tree and Wet tree system

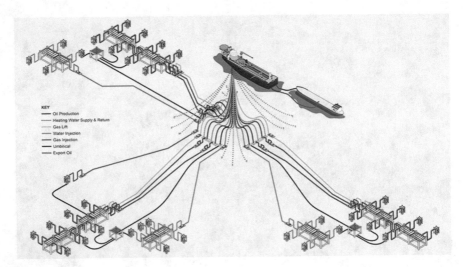

Fig. 7.8 Tieback field Architecture

- Low cost hull forms are feasible
- Simplified riser/vessel interfaces.

Dry tree systems: Dry tree systems are the optimum option to the subsea well cluster architecture, where, the surface well architectures show easy vertical access to the wells for future intervention activities. Risers for dry completion systems can be either single casing, dual casing, combo risers, or tubing risers and might contain a split tree. The riser tensioning system as well offers some alternatives such as active hydropneumatic tensioners, air cans, locked-off risers, or king-post tensioning mechanism.

Chapter 8
CAPEX and OPEX Expenditures

This chapter is discussing the expenditures associated to all investment (CAPEX) that is made in advance, and the operational expenses (OPEX) that will occur from the first day startup the production until the abandonment day. Roughly numbers will be presented for these expenses in deepwater projects and will see what is the key assumptions must be considered when keying these values to any project. Accessible records that cover these values will be revealed together with the approximation costs and methodology.

Typically, oil and gas resources take inputs and turn them into products that make profits. Figure 8.1 is an image of the profits generation process arising from the capex and OPEX investment in an oil and gas asset.

8.1 Surface Facilities and Subsea Costs (Capex)

The term CAPEX covers the costs related to the drilling wells and constructing the surface treatment facilities either for onshore or offshore platforms such as, FPSOs, SPARs, semi-submersibles, includes also dry and wet trees, subsea manifolds, and pipelines.

Normally, the operations companies use a specific benchmarks and internal costs for all field operations costs. This book aims to provide roughly figures which can help the owners of the projects to estimate the required operations costs for their projects, also to comparing that with the used benchmarks.

8.2 Drilling Costs

There are to mean parameters reflecting the costs of drilling expenses which are the well depth and the daily rig rate. Normally, the cost of the daily rig rate is varying according to the drilling depth, water depth, rig type, distance from shore.

© The Author(s), under exclusive license to Springer Nature Switzerland AG 2020
T. A. O. Ganat, *Technical Guidance for Petroleum Exploration and Production Plans*,
SpringerBriefs in Applied Sciences and Technology,
https://doi.org/10.1007/978-3-030-45250-6_8

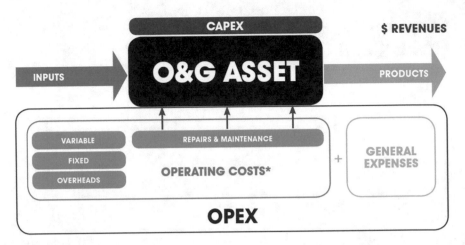

Fig. 8.1 Schematic OF revenue generation process in an oil and gas asset

Usually, for onshore drilling operations, the drilling costs might be reached 100,000$/d, however, for deepwater offshore the drilling costs can vary between 600,000$/d to 800,000$/d (reference year 2010). The total number of days required to drill the well is a function of total depth intend to drill. For example, for depths up to 20,000 ft, the total drilling days can vary between 70 days to 80 days and for larger depths up to 32,000 ft, a maximum drilling days might reach 150 days. The data in Table 8.1, can be used to estimate the total cost of the dry hole well. For the estimation of the appraisal and development wells, some extra costs need to be added such as well testing and well completion costs. For instance, if a reservoir discovery has an oil initial in place is 120 MMbbl of oil, one exploratory well and one appraisal well need to be drilled based on the total area of the reservoir. After the confirm the commerciality of the project, 13 development wells will be drilled. The total cost of one exploratory well will vary from 55 to 88 MM$, and for one appraisal well the cost varies also from 55 to 88 MM$ plus well testing costs (±15 days extra) about 5MM$. So the total appraisal well cost is between 60 MM$ to 93 MM$.

In Chap. 7, Fig. 7.5 shows the topsides and hull costs for semisubmersible production units which can be considered as linear relation. Where the X-axis is unit capacity in thousands of barrels per day and the Y-axis the cost of hull and the topside (fabrication 1 installation) in MM$. If the production capacity of the project can

Table 8.1 Drilling Costs as a Function of Rig Rate (2010) and Drilling Time

Rig Rate (MM$/day)	Total depth (ft)	Total drilling days	Total drilling cost/well (MM$)
500–800	20,000	70–80	35–64
500–800	26,000	110	55–88
500–800	32,000	150	75–120

accommodate 35,000 bbl/d, then trend line equations can be generated and utilized to the estimate the costs of the hull and topsides. From the plot the estimated hull cost will be about 410 MM$ and topside costs about 560 MM$, with a total of 970 MM$. We may need to add some contingency costs around 25%, so the total cost will be about 1,213 MM$.

For estimation the subsea costs and export pipelines costs, for 13 development wells, two manifolds need to be installed (one manifold will tie-in with 6 wells and another manifold will tie-in with 7 wells) plus inter-field flow lines and risers. Normally these costs are estimated using pervious projects figures and/or from suppliers. If we considered the following assumptions:

- Well templates and manifolds 34 MM$/well.
- Four production risers at 0.35 MM$/(1000 ft).
- One export riser at 0.28 MM$/(1000 ft).
- Flow lines and export pipelines at 0.23 and 0.16 MM$/(miles). Let us assume flow lines length 5 miles and export pipeline length 30 miles.

So, the cost of the templates and manifolds for 13 wells is 442 MM$, and 4 production risers at water depth 5500 ft is 7.7 MM$. The cost of the export riser is at water depth 5500 ft is 1.54 MM$, and the cost of the flow lines for 30 miles is 6.9 MM$.

8.3 Operating Expenditure (OPEX)

The other term usually used in cash flow, operating expenditure (OPEX), comprises of the periodic costs of the company. It can also cover worker's costs and facility expenditures such as maintenance, rent…etc.

For example, a well's production life is varying from 4 to 30 years. During this period, both the planned operations and maintenance (O&M) costs and the unplanned O&M expenditures are required to compute life cycle expenditures during the production life. Therefore, OPEX includes the planned recompletions cost which is the intervention rig spread cost multiplied by the expected recompletion time for every single recompletion. The timing and number of planned recompletions are commonly reliant on the field development plan.

Comprehensive OPEX cost model can be established by research and discussions. The worksheet model can be used to assist the project operator to get a clear idea of the cost effect by modifying the values of the variable cost and directly seeing the cost effect of the changes made. This approach helps the operator to identify the effect and cost sensitivity of any operating variable. The OPEX cost model can be applying as an approach to find the operating cost ranges for field development.

The OPEX are very complex and hard to guess by simply using factors or charts. Usually it's a function of the maximum production rate and Pipeline tariff. Table 8.2 shows an example of a tentative summary of OPEX costs per barrel of oil equivalent for the years applicable on plateau.

Table 8.2 Total OPEX Estimation

Facility type	No tariff	With tariff
Standalone	8–15 $/boe	12–25 $/boe
Tiebacks	5–10 $/boe	8–13 $/boe

Table 8.3 Project timing period starting from exploratory well to first oil

Facility type	Time (Years)
Standalone	1 year—Exploratory well, 2 years—Appraisal well 3–7 years (facilities 1 subsea)
Tiebacks	As above Up to 3 years for the facility tieback

So, for standalone project with tariff, the total OPEX for the field life with total production rate of 125 MMbbl as a minimum is 1,500 MM$ and as a maximum is 3,125 MM$ for the 9 years of production on plateau. This total cost must be prorated uniformly to the production period for the remaining years.

For the plateau rate, the yearly OPEX would be 1,500 MM$/9 years or 3,125 MM$/9 years and prorated for the rest of the years. At the years where their production rates are 25% lower than the plateau rate, a minimum 25% OPEX is recommended for these years to take into consideration the maintenance costs and other costs that are not a direct related to the production process.

8.4   Project Timing and Input to Economic Analysis

At this stage, all the inputs have been calculated which are needed for our economic analysis and to the end of our volume to value process.

The timing of the development plan is critical in relation to all the other inputs, on the other hand, it is flexible reliant on how the operator desires to drilling the wells and investing ahead for the construction of facilities.

Preferably, to drill at least one well per year and assign a period of time in which to manufacture the planned field facilities and to have them in place. Table 8.3 displays the total project time starting from exploration activities to first production. Typically, the best preferred scenario, this period would not be less than 5 years. Normally, for a giant project, this period may extend to 10 years or more.

Chapter 9
Case Studies Evaluations

This chapter will show some case studies from offshore deepwater oil and gas fields at several depths as well as different field development options such as standalone hosts and tie-backs.

9.1 Case Study-1: Small Heavy Oil Accumulation Discovery

The offshore discovery is a small accumulation of heavy oil, low dissolved gas (Fig. 9.1). No exploratory wells available, and all data is provided from seismic. Table 9.1 shows the input data available from the play.

9.1.1 Primary Estimation of the Recovery Factor

The engineers need to find the optimum analogy information, either locally or internationally to estimate the expected oil recovery factor for their hydrocarbon accumulation. Figure 9.2 shows the recovery factor from different depositional age at different depths. From the plot the expected recovery factor at 25,000 ft reservoir oil depth is 25% for a Paleogene prospect. So, the expected heavy oil recoverable volume will be 125 MMbbl.

9.1.2 Rock/Fluid Properties and Well Count Estimation

In the next stage, engineers need to estimate the total number of wells required along with their initial forecasting flow rate. Therefore, reservoir rock and fluid properties need to be estimated. If there is no available data from the same prospect, analog

© The Author(s), under exclusive license to Springer Nature Switzerland AG 2020
T. A. O. Ganat, *Technical Guidance for Petroleum Exploration and Production Plans*,
SpringerBriefs in Applied Sciences and Technology,
https://doi.org/10.1007/978-3-030-45250-6_9

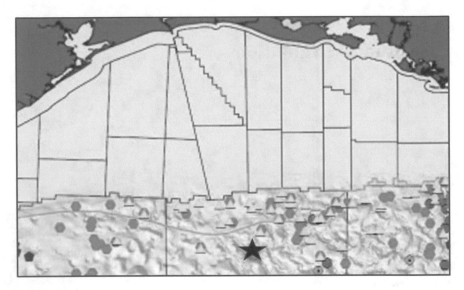

Fig. 9.1 Prospect location map

Table 9.1 Prospect input data available

Parameter	Value
OIIP (MMbbl)	500
Water depth (ft)	5,500
Deposition age	Paleogene
Area (acres)	3,420
Net pay Thickness (ft)	300
Total depth (ft)	25,000

information can be used to estimate the well potential. Figure 9.3 shows porosity variation with depth at different deposition ages. At reservoir depth 25,000 ft the estimated average porosity value will be within the range from 10% to 20% for Oligocene (Upper Paleogene). By using this range, permeability can be estimated from Fig. 9.4 which is within the range of $K = 5$–75 mD.

Still need to estimate some of fluid parameters required to calculate the expected well flow rate such as viscosity and API grade for the oil. Most of the Paleogene discoveries have viscosity around 10 cP and API range from 20 API to 22 API as seen in Fig. 9.5.

Once permeability (K), net thickness (H) and Viscosity (μ) are obtained, the next step is to calculate the index (kh/μ) to estimate the total recovery per well along with the total number of development wells required. As $K = 55$ mD (min) and 75 mD (max), $H = 5300$ ft and $\mu = 510$ cP, the index (min) $= 55 \times 300/10 = 5150$ (mD ft/cP) and Index (max) $= 575 \times 300/10 = 52250$ (mD ft/cP).

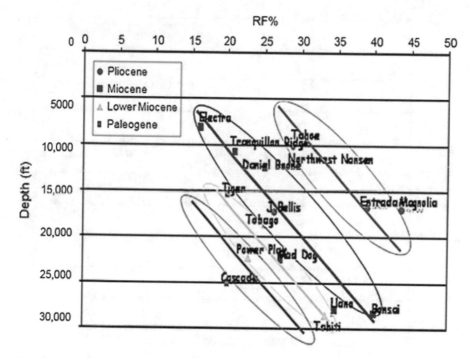

Fig. 9.2 Example of a trend deriving using analogs and grouping parameters

Figure 9.6, shows the index numbers (150 and 2,250) are located at the lower index scale, with very low oil recovery per well. As a rule of thumb, the estimated oil recovery per well is acceptable for such API oil gravity and such reservoir permeability. The plot shows that the ultimate recovery per well (UR/well) is within the range between 2 MMbbl to 10 MMbbl. Therefore, the total number of development oil wells at low UR is 125 MMbbl/2 MMbbl = 563 wells and a high UR is 125 MMbbl/10 MMbbl = 513 wells. Accordingly, at low UR the project is not economical with high number of wells need to be drilled. In such a case, high UR will be selected as the best option and its more economical case.

9.1.3 Well Initial Rates and Notional Production Profile

The next step in the development plan is to estimate the type of the surface treatment facilities required and their capacities. For this reason, the maximum oil production rate from all development wells need to be calculated and create a production curve with time.

By using the analogues information from the same region producing heavy oil with average API between 22 to 24API (Fig. 9.7). As KH = 575 × 300 = 522,500 mD ft, the average oil rate per well is about 2,300 bbl/d. Therefore, the total oil

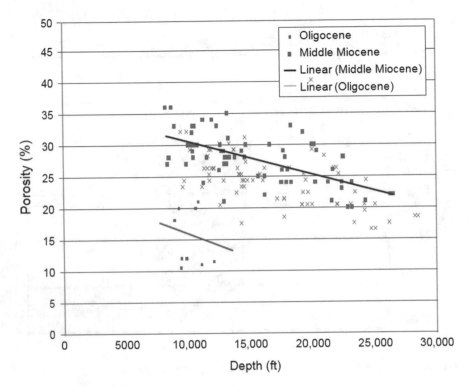

Fig. 9.3 Trend for porosity versus depth

production from the 13 wells at 2,300 bbl/d will be 29,900 bbl/d. The next step is to generate production forecasting profile for all the field. Figure 9.8 shows the first scenario of production curve versus time that produces a plateau rate of 29,900 bbl/d for 9 years and a total recoverable reserves of 125 MMbbl.

The production profile at the first year (2015) is 15,000 bbl/day which call ramp up period due to small amount of wells (6 oil wells) already drilled before production day and tied to the production flow line at the first oil date followed by the remaining drilled oil wells (7 oil wells) in the next year (2016). The potential of the total oil production from the 13 wells was sustained for 9 years as plateau period and then the production drops down at constant production decline rate until the economic rate (end of the production life of the field).

9.1.4 Estimation the Capex and Opex Costs

To execute the development plan, it is required to know the total project cost to drill the 13 development wells, wells completions, construct the surface treatment facilities, and oil evacuation system along with the expenditure costs.

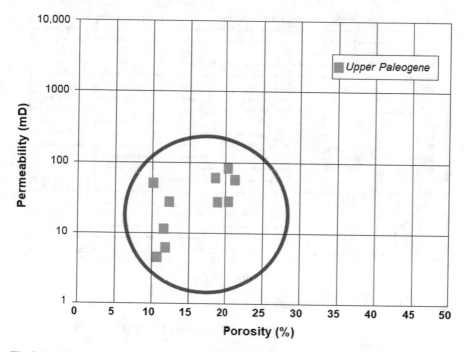

Fig. 9.4 Permeability for Oligocene play from a database with analog information

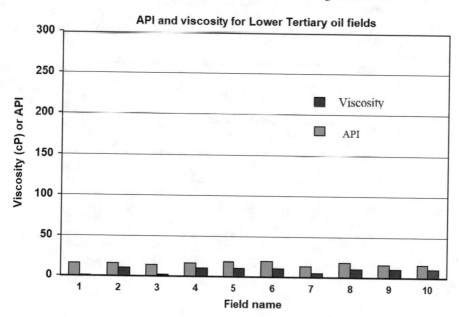

Fig. 9.5 Viscosity for fields in the Lower Tertiary

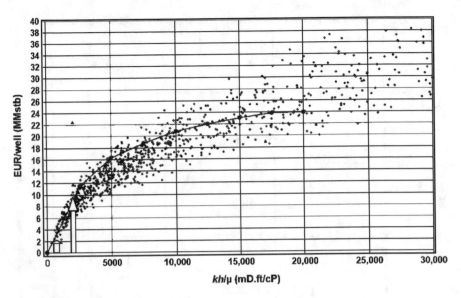

Fig. 9.6 Estimation number of oil wells using reservoir rock and fluid properties

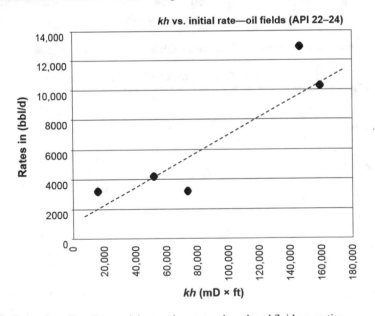

Fig. 9.7 Estimation oil well potential rate using reservoir rock and fluid properties

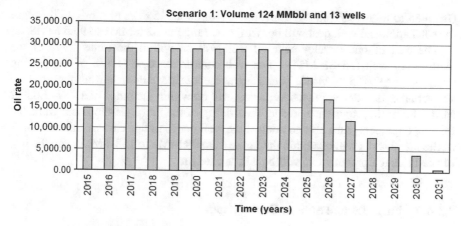

Fig. 9.8 Production forecasting profile

9.1.4.1 Drilling Costs

Normally, the drilling costs are reliant on the reservoir depth and the daily rig rate cost. Typically, the daily rig rate cost is depending on the rig type, water depth (offshore), drilling depth, and the location distance from shore. For onshore drilling rig, normally it is about 100,000$/day, and for deepwater offshore, it is varying between 500,000$/day and 800,000$/day. The total number of drilling days is reliant on the well depth. Usually, at depth up to 20,000 ft, the drilling operation days may extend between 70 days and 80 days and for ultra-depths like 32,000 ft and more, a maximum drilling operations days might reach 150 days.

Table 9.2 shows the drilling costs as a function of daily rig rate and drilling time to estimate the dry hole cost.

The range of the well cost shown in the table is the cost of the exploratory well from the analogues. As discussed in the Chap. 8, the appraisal and developments wells costs are similar to the cost of the exploratory well plus some extra costs must be added such as well testing costs and well completion costs. As the accumulation oil of this field is 125 MMbbl of oil, then the proposed plan is to drill 1 exploratory well and 1 appraisal well and 13 development wells.

From the given data in the Table 9.2, the cost of 1 exploratory well is within the range of 55–88 MM$, and the cost of 1 appraisal well will be within the same range

Table 9.2 Drilling costs as a function of rig rate and drilling time

Rig rate (MM$/day)	Depth (ft)	Drilling days	Well cost (MM$)
500–800	20,000	70–80	35–64
500–800	26,000	110	55–88
500–800	32,000	150	75–120

(from 55 to 88 MM$) plus well testing cost of about 15 days or more. In this case the total appraisal well cost will be within the range of 62.5 MM$ to 95.5 MM$.

The cost of one development oil well will be within the same range of the exploratory well (55–88 MM$) plus well completion cost (180% of the drilling cost) totaling 99–158.4 MM$. Therefore, as the total development wells required are 13 well, the total development wells cost will be within the range of 1,287–2,059.2 MM$. From the experience, during the drilling of the first 2 development wells the drilling engineers learn how to control the well costs, in this case assume the costs of the first 2 wells will be within the range from 99 MM$ to 158.4 MM$ and the rest of the 11 development wells will be a little cheaper.

9.1.4.2   Facilities and Subsea Costs (Capex)

Figure 9.9 shows that topsides and hull for semisubmersible production units which can be seen its reliant on its total facility capacity. The plot shows almost liner relationship. The numbers obtained from the figure, may not be very accurate, so more care must be taking once used such numbers.

The X-axis represents facility capacity and the Y-axis is the Topside and Hull cost (fabrication plus installation). As far as the total production rate from 13 well is 29,900 bbl/day, the surface facilities need to designed to accommodate 30,000 bbl/d. By using the trend equations in Fig. 9.9, the Hull and Topside costs for 30 Mbbl/d, is about 400 MM$ and 552 MM$ respectively, with a total of 952 MM$. Some contingency costs might be added here around 25%.

Now, the subsea costs and export pipelines costs need to be estimated. Typically, for 13 development wells, two manifolds are required to handle all the wells (one manifold will tie-in with 6 wells and another manifold will tie-in with 7 wells) plus inter field flow lines and risers. Now let's assume some factors to calculate costs:

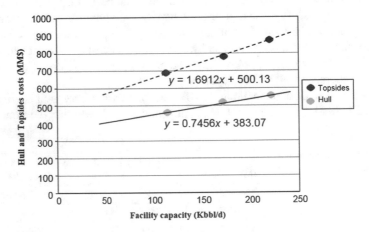

Fig. 9.9 Hull and topsides Capex estimation versus facility capacity

- Well templates and manifolds 34 MM$/well.
- SCR Production riser 2 risers per manifold 4 risers at 0.35 MM$/(1000 ft).
- 1 SCR Export riser at 0.28 MM$/(1000 ft).
- Flow lines and export pipelines at 0.23 and 0.16 MM$/(miles). Flow lines length is 5 miles and export pipeline length is 30 miles.

These are estimated factors as provided by suppliers. Templates and manifolds cost at 34 MM$/well for 13 wells is 442 MM$, and for production risers at water depth 5500 ft is 924 MM$. Export riser at 0.28 MM$/(1000 ft) and at water depth 5500 ft is 15.4 MM$. For 5 miles' flow line length and for 30 miles' export pipeline cost is 48 MM$.

9.1.4.3   Operating Costs (Opex)

Table 9.3 shows OPEX estimation for the years on plateau or peak rates. As this project is standalone with tariff, then total expected OPEX for the field life at low cost (12$/boe) for recoverable oil volume of 125 MMbbl is 1,500 MM$ and for high cost (25$/boe) is 3,125 MM$ for the 9 years of production on plateau period. This total cost must be prorated uniformly to production for the rest of the field life.

As mentioned in Chap. 8, at the years where their production rates are 25% less than the plateau production rate, a minimum 25% OPEX is suggested for these years to consider the maintenance costs and other costs that may not related to the production process directly.

9.1.5   Project Timing and Input to Economic Analysis

Most of the data required for economic analysis are estimated. Project timing data need to be obtained from analogues as shown in Table 9.4.

Table 9.3 Total OPEX estimation

Facility type	No tariff	With tariff
Standalone	8–15$/boe	12–25$/boe
Tie backs	5–10$/boe	8–13$/boe

Table 9.4 Project timing

Facility type	Time (Years)
Standalone	1 year Exploratory well/2 years Appraisal well 3 to 7 years (facilities 1 subsea)
Tiebacks	1 year Exploratory well/2 years Appraisal well 3 to 7 years (facilities 1 subsea) Up to 3 years for the facility tieback

The project timing is critical same the other inputs, however it is flexible depending on how the operator needs to drill the wells and invest in advance on the fabrication of facilities.

As all the required information are gathered and estimated, net present value (NPV) need to be calculated for the project to evaluate the opportunity considering all the economic parameters. Figure 9.10 summarizes the inputs that any economic package will require, basically the revenues (volumes) and the expenditures. Note that in the Fig. 9.10, there are min and max numbers for the costs. The user may choose to pick an average number before running the economics.

9.2   Case Study-2: Big Light Oil Accumulation Discovery

The prospect is located at deepwater south American region at water depth 5,500 ft (Fig. 9.11). The discovery found in Lower Miocene age. Table 9.5 shows the input data available.

9.2.1   Primary Estimation of the Recovery Factor

Once again, the first stage to start up the project evaluation is to estimate the recover factor for the oil in place. By using the analog information in Fig. 9.2, the expect recovery factor is around 19% for a lower Miocene prospect with reservoirs at depth of 15,000 ft. Therefore, the recoverable volume from the initial oil in place will be 190 MMbbl.

9.2.2   Rock/Fluid Properties and Well Count Estimation

At this step, the number of wells and their initial rate must be derived. Therefore, some reservoir rock and fluid properties need to be estimated.

Figure 9.12 shows porosity variation with depth at different deposition environment. At reservoir depth 15,000 ft the estimated average porosity value will be 28% for Miocene. By using this porosity, permeability can be estimated from Fig. 9.13 which is within the range of 500–1000 mD.

Once permeability, net thickness and Viscosity are obtained, the next step is to calculate the index (kh/μ) to estimate the total recovery per well along with the total number of development wells required. As K = 5500 mD (min) and 1,000 mD (max), H = 5200 ft and μ = 52 cP, the index (min) = 5500 × 200/2 = 550,000 (mD ft/cP) and Index (max) = 51,000 × 200/2 = 5100,000 (mD ft/cP).

Date	Oil Production rate (stb/day)	Gas Production rate (MMscf/day)	Exploration Drilling (MM$) min-max	Appraisal Drilling (MM$) min-max	Development Drilling (MM$) min-max	Capex (MM$) Facility	Capex (MM$) Subsea	Capex (MM$) pipeline	Abandonment (MM$)	Opex (MM$)	Comments
2010			55-88								1 Exploratory
2011				62.5-100	(99-158) × 1	255.6					1 Appraisal+1 well
2012					(99-158) × 3	255.6					3 wells
2013					(99-158) × 3	255.6	147.3333	924			3 wells
2014					(99-158) × 3	255.6	147.3333	15.4			3 wells
2015	14,300.00				(99-158) × 3	255.6	147.3333	48			3 wells
2016	28,600.00									33.33-173.61	
2017	28,600.00									66.66-347.22	
2018	28,600.00									66.66-347.22	
2019	28,600.00									66.66-347.22	
2020	28,600.00									66.66-347.22	
2021	28,600.00									66.66-347.22	
2022	28,600.00									66.66-347.22	
2023	28,600.00									66.66-347.22	
2024	22,600.00									83.33-173.61	
2025	17,000.00									41.67-86.81	
2026	12,000.00									41.67-86.81	
2027	8,000.00									41.67-86.81	
2028	6,000.00									41.67-86.81	
2029	4,000.00									41.67-86.81	
2030	1,000.00									41.67-86.81	
2031									127.60		10% of Total Facility Capex
Total	volume = 125 MMbbl		55-88	62.5-100	1287-2054	1276	442		127.60		

Fig. 9.10 Economic analysis inputs

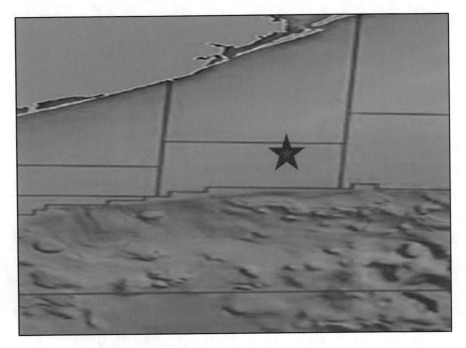

Fig. 9.11   Location Map of a light oil prospect in deepwater

Table 9.5 Prospect input
data available

Parameter	Value
OIIP (MMbbl)	1,000
Water depth (ft)	5,500
Deposition age	Paleogene
Area (acres)	1,200
Net pay Thickness (ft)	200
Total depth (ft)	15,000

So, after determined the recovery factor, porosity and permeability ranges using analogues information, API oil gravity need to be estimated. From analogues information, the expected oil gravity is API 32 and oil viscosity is 2 cP.

Figure 9.14, shows the index numbers (50,000–100,000) are located at the upper limit scale. The UR/well in some of the produced wells at the analogues area recovered about 30 MMbbl/well. Therefore, in this prospect the same high oil recovery per well (UR/well = 530 MM bbl) will be expected. As a rule of thumb, the estimated oil recovery per well is acceptable for such API oil gravity with high reservoir permeability in deepwater prospects. Therefore, the total number of development oil wells at high UR will be 190 MMbbl/30 MMbbl = 7 wells.

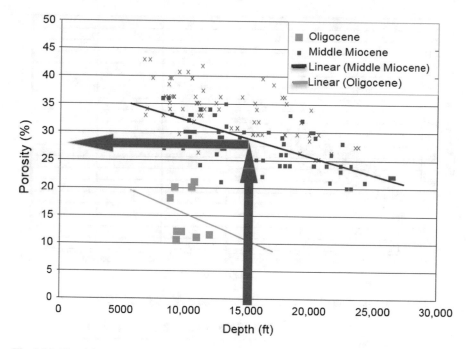

Fig. 9.12 Trend for porosity versus depth from analogy

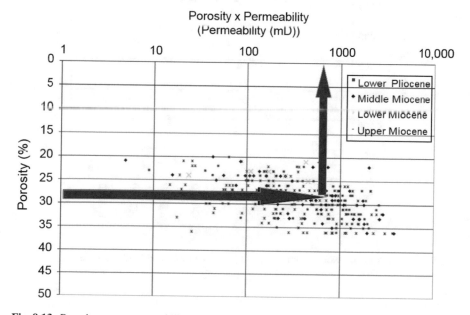

Fig. 9.13 Porosity versus permeability trend from analogy

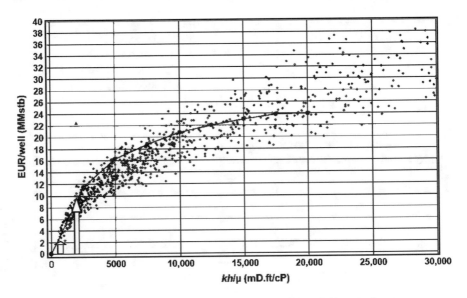

Fig. 9.14 Estimation number of oil wells using reservoir rock and fluid properties

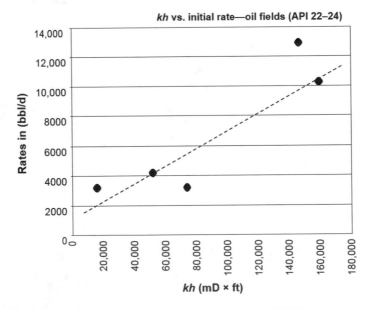

Fig. 9.15 Estimation oil well potential rate using reservoir rock and fluid properties

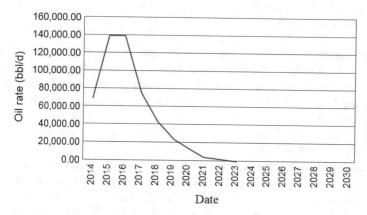

Fig. 9.16 Production forecasting profile

9.2.3 Well Initial Rates and Notional Production Profile

Figure 9.15 was used to estimate the oil rate per well. This Figure is derived from analogues (oil fields with API averaging around 24 API). As our fluid, we will have a higher API (32) and lower viscosity to be conservative.

At low permeability range, KH = 5500 × 200 = 5100,000 mD. ft, the average oil rate is about 7,500 bbl/d. And at high permeability, KH = 51,000 × 200 = 5200,000 mD ft, the average oil rate is about 12,000 bbl/d or more. For a fluid of 32 API these rates could be 20,000 bbl/d.

Therefore, the total oil production from the 7 wells at 7,500 bbl/d will be 52,500 bbl/d, and at 20,000 bbl/d will be 140,000 bb/d. The next step is to generate production forecasting profile for all the field. Figure 9.16 shows the first scenario of production curve versus time that produces a plateau rate of 140,000 bbl/d for 7 years and a total recoverable reserves of 190 MMbbl.

The production profile at the first year (2014) is 60,000 bbl/day which call ramp up period due to small amount of wells (3 oil wells) already drilled before production day and tied to the production flow line at the first oil date followed by the remaining drilled oil wells (4 oil wells) in the next year (2015). The potential of the total oil production from the 7 wells was sustained for 3 years as plateau period and then the production drops down at constant production decline rate until the economic rate.

9.2.4 Estimation the Capex and Opex Costs

To execute the development plan, it is required to know the total project cost to drill the 7 development wells, wells completions, construct the surface treatment facilities, and oil evacuation system along with the expenditure costs.

9.2.4.1 Drilling Costs

As mentioned early, the drilling costs are reliant on the reservoir depth and the daily rig rate cost. Normally, the daily rig rate cost is depending on the rig type, water depth, drilling depth, and the location distance from shore. Drilling rig cost for deepwater, it is varying between 500,000$/day to 800,000$/day. The total number of drilling days is reliant on the well depth.

Table 9.2 shows the drilling costs as a function of daily rig rate and drilling time to estimate the dry hole cost.

The range of the well cost shown in the table is the cost of the exploratory well from the analogues. As discussed in the previous case study, the appraisal and developments wells costs are similar to the cost of the exploratory well plus some extra costs must be added such as well testing costs and well completion costs. The accumulation oil of this field is 190 MMbbl of oil, so to recover this accumulation, the proposed plan is to drill 1 exploratory well and as the volumes are between 100 MMbbl and 200 MMbbl one appraisal well need to be drilled and 7 development wells.

From the given data in the Table 9.2, the cost of 1 exploratory well will be within the range of 36 MM$ to 64 MM$, and the cost of 1 appraisal well will be within the same range (from 35 to 64 MM$) plus well testing cost of about 15 days or more. In this case the total appraisal well cost will be within the range of 42.5 MM$ to 76 MM$.

The cost of one development oil well will be within the same range of the exploratory well (35–64 MM$) plus well completion cost (180% of the drilling cost) totaling 63–115 MM$. Therefore, as the total development wells required are 7 well, the total development wells cost will be within the range of 441–805 MM$. From the experience, during the drilling of the first 2 development wells the drilling engineers learn how to control the well costs, in this case assume the costs of the first 2 wells will be within the range from 63 MM$ to 115 MM$ and the rest of the 5 development wells will be a little cheaper.

9.2.4.2 Facilities and Subsea Costs (Capex)

Figure 9.17 shows that topsides and hull for semisubmersible production units which can be seen its reliant on its total facility capacity. The plot shows almost liner relationship. The numbers obtained from the figure, may not be very accurate, so more care must be taking once used such numbers.

If we select standalone project than the surface facilities need to be designed to accommodate 140,000 bbl/d. By using the trend equations in Fig. 9.17, the Hull and Topside costs for 140,000 bbl/d, is about 490 MM$ and 750 MM$ respectively, with a total of 952 MM$. Some contingency costs might be added here around 25%.

But in this case study, we will need a hub, however let's explore the possibility for a subsea tieback to the nearby existing field where there are already some prospects being developed. So, the only things needed are subsea and pipe line costs. Will use

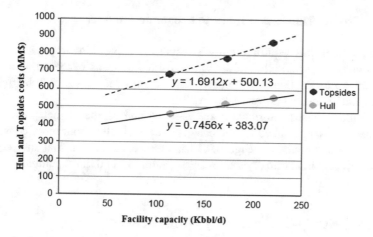

Fig. 9.17 Hull and topsides Capex estimation versus facility capacity

one manifold only for 7 wells, plus export pipeline to the platform and the riser. Costs should look like:

- Well templates and manifolds 34 MM$/well
- SCR production riser at 0.35 MM$/(1000 ft)
- Export pipelines at 0.16 MM$/(miles). Let's assume that the export pipeline length equals 5 miles.

These are estimated factors as provided by suppliers. Templates and manifolds cost at 34 MM$/well for 7 wells is 238 MM$, and for production risers at water depth 5500 ft is 28.9 MM$. Export pipelines for 5 miles at 0.16 MM$/(1000 ft) is 6.4 MM$. In this case, a bigger pipe diameter need to handle 140,000 bbl/d. Therefore, the cost of the riser may go up to 5 MM$.

9.2.4.3 Operating Costs (Opex)

Table 9.3 shows OPEX estimation for the years on plateau or peak rates. As this project is standalone with tariff, then total expected OPEX for the field life at low cost (8$/boe) for recoverable oil volume of 190 MMbbl is 1,520 MM$ and for high cost (13$/boe) is 2,470 MM$ for the 3 years of production on plateau period. This total cost must be prorated uniformly to production for the rest of the field life.

For the peak rates on plateau our yearly OPEX would be 50% which is 1,520/3 = 5253 MM$ or at Maximum tariff which is 2,470/3 = 5411.66 MM$ and prorated for the rest of the production years. Again, a minimum of 25% Opex is recommended for the years where the production is lower than 25% of peak rate.

9.2.4.4   Project Timing and Input to Economic Analysis

A subsea tieback will require roughly 3 years from exploration to production time. During this time period all 7 wells will be drilled (2 wells per year). Our project might start up production in 2015 assuming the exploratory well will be drilled in 2011 and the appraisal one year later in 2012.

As all the required information are gathered and estimated, net present value (NPV) need to be calculated for the project to evaluate the opportunity considering all the economic parameters. Figure 9.18 summarizes the inputs that any economic package will require, basically the revenues (volumes) and the expenditures.

9.3   Case Study-3: Small Dry Gas Discovery

The offshore discovery is a dry gas accumulation located on offshore at deepwater (Fig. 9.19). there is one exploratory well available, and all data is provided from seismic. Table 9.6 shows the input data available from the prospect.

9.3.1   Primary Estimation of the Recovery Factor

Again, the first stage to start up the project evaluation is to estimate the recover factor for the dry gas in place. By using the analogues information in Fig. 9.20, the expect recovery factor is around 60% for the Pliocene play with reservoirs at depth of 17,000 ft. Therefore, the recoverable volume from the initial dry gas in place will be 300 bcf.

9.3.2   Rock/Fluid Properties and Well Count Estimation

At this step, the number of wells and their initial rate must be derived. Therefore, some reservoir rock and fluid properties need to be estimated.

Figure 9.21 shows porosity variation with depth at different deposition environment. At reservoir depth 17,000 ft the estimated average porosity value will be 15% for Pliocene. By using this porosity, permeability can be estimated from Fig. 9.22 which is within the range of 10 mD to 100 mD.

Up to now, recovery factor, porosity and permeability ranges using analog information already estimated. By using permeability and net thickness and also assuming that the gas has a very low viscosity (0.1 cP), recovery per well and the total number of producers can be estimated.

At low permeability range, KH = 510 × 200/0.1 = 520,000 mD ft, and at high permeability, KH = 5100 × 200/0.1 = 5200,000 mD ft. The UR/well would be very

Date	Oil Production rate stb/day	Gas Production rate MMscf/day	Exploration Drilling MM$ min-max	Appraisal Drilling MM$ min-max	Development Drilling MM$ min-max	Capex MM$ Facility	Capex MM$ Subsea	Capex MM$ pipeline	Abandonment MM$	Opex MM$	Comments
2010											
2011			35-64								
2012				42.5-76							1 Exploratory
2013					(63-115) × 1						1 Appraisal + 1 well
2014					(63-115) × 2		119	28.9			2 wells
2015	70,000.00				(63-115) × 2		119	6.4			2 wells
2016	140,000.00									126-205	
2017	140,000.00									253-411	
2018	77,000.00									253-411	
2019	45,000.00									126-205	
2020	25,000.00									63-102	
2021	15,000.00									63-102	
2022	5,000.00									63-102	
2023	2,500.00									63-102	
2024	1,000.00									63-102	
2025										63-102	
2026											
2027											
2028											
2029											
2030											
2031											
2032											
Total			35-64	42.5-76	441-805		238	35.3			

Fig. 9.18 Economic analysis inputs

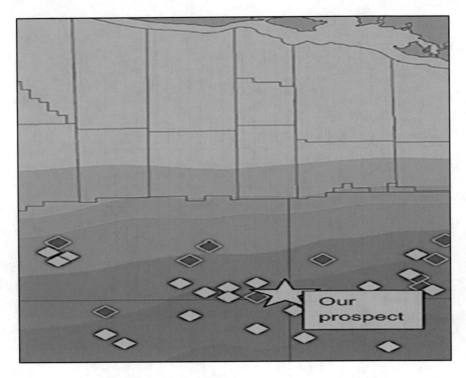

Fig. 9.19 Location of a dry gas prospect in deepwater

Table 9.6 Prospect input data available

Parameter	Value
GIIP (bcf)	500
Water depth (ft)	6,900
Deposition age	Pliocene
Area (acres)	2000
Net pay Thickness (ft)	200
Total depth (ft)	17,000

high more than 25 MMbbl. As its gas wells, the equivalent UR/well is 150 bcf/well. The total number of gas wells required is 300 bcf/150 bcf = 52 gas wells.

9.3.3   Well Initial Rates and Notional Production Profile

Figure 9.24 was used to estimate the oil rate per well. This figure is derived from analogues (oil fields with API averaging ranges between 22 API and 24 API). To

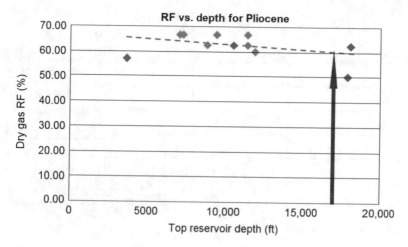

Fig. 9.20 Example of a trend deriving using analogs, RF versus top reservoir depth

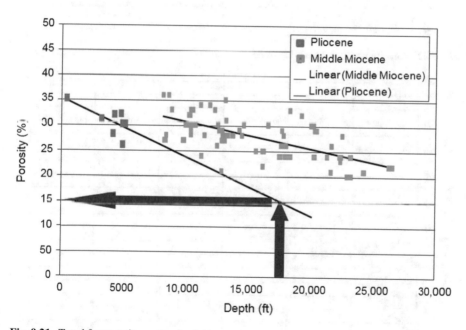

Fig. 9.21 Trend for porosity versus depth

apply it for gas, we need to assume a viscosity for gas of 0.1 cP and that the rates will be in BOE (barrels of oil equivalent).

At low permeability range, KH = 5100 × 200 = 52,000 mD ft, the average oil rate is about 2,000 bbl/d equivalent to 2 MM scf/d. And at high permeability, KH = 5100 × 200 = 52,000 mD ft, the average oil rate is about 12,000 bbl/d equivalent to 72 MMscf/d. In this project, 10,000 bbl/d equivalent to 60 MMscf/d was proposed

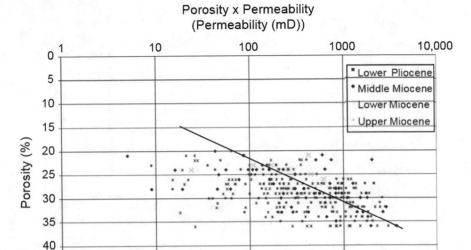

Fig. 9.22 Porosity versus permeability trend

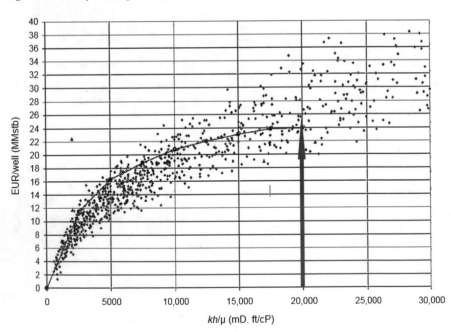

Fig. 9.23 Estimation of number of wells based on rock and fluid propertie

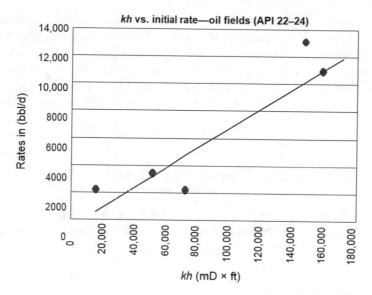

Fig. 9.24 Estimation of well rates based on rock and fluid properties

as average flow rate for the dry gas play. Therefore, the total gas production from the 2 wells at 60 MMscf/d will be 120 MMscf/d.

The next step is to generate gas production forecasting profile for all the field. Figure 9.25 shows the first scenario of production curve versus time that produces a plateau rate of 120,000 MMscf/d for 4 years and a total recoverable reserves of 300 bcf.

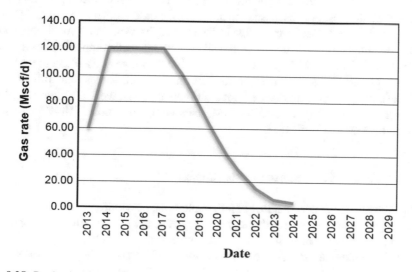

Fig. 9.25   Production forecasting profile

The production profile at the first year (2013) is 60,000 bbl/day which call ramp up period due to small amount of wells (1 gas well) already drilled before production day and tied to the production flow line at the first gas production date followed by the second drilled gas well (1 gas well) in the next year (2014). The potential of the total gas production from the 2 wells was sustained for 4 years as plateau period and then the production drops down at constant production decline rate until the economic rate.

9.3.4 Estimation the Capex and Opex Costs

To execute the development plan, it is required to know the total project cost to drill the 2 development wells, wells completions, construct the surface treatment facilities, and gas evacuation system along with the expenditure costs.

9.3.4.1 Drilling Costs

The drilling costs are reliant on the reservoir depth and the daily rig rate cost. Normally, the daily rig rate cost is depending on the rig type, water depth, drilling depth, and the location distance from shore. Drilling rig cost for deepwater, it is varying between 500,000$/day to 800,000$/day. The total number of drilling days is reliant on the well depth.

Table 9.2 shows the drilling costs as a function of daily rig rate and drilling time to estimate the dry hole cost.

The range of the well cost shown in the table is the cost of the exploratory well from the analogues. As discussed in the previous case studies, the appraisal and developments wells costs are similar to the cost of the exploratory well plus some extra costs must be added such as well testing costs and well completion costs. For a depth of 17,000 ft, the drilling costs will be within the range of 35 MM$ to 64 MM$ for the exploratory well. As the volumes are small which is less than 100 MMbbl (300 Bcf), the project team decided not to drill an appraisal well. So, a development well will cost from 35 MM$ to 64 MM$ plus completion costs (180%) which is between 63 MM$ to 115 MM$.

9.3.4.2 Facilities and Subsea Costs (Capex)

In this study, we may need a hub, but let's explore the possibility for a subsea tieback to the neighboring area where there are already several projects being developed. In this case, we will only need subsea and pipeline costs. For 2 wells, we will need only one manifold plus export pipeline to the platform and the riser. The costs should look like:

- Well templates and manifolds 34 MM$/well
- SCR production riser at 0.35 MM$/(1000 ft)
- Export pipelines at 0.16 MM$/(miles). Let's assume that the export pipeline length equals 5 miles.

These are estimated factors as provided by suppliers. Templates and manifolds cost at 34 MM$/well for 2 wells is 68 MM$, and for production risers at water depth 6,900 ft is 36.23 MM$. Export pipelines for 5 miles at 0.16 MM$/(1000 ft) is 6.4 MM$.

9.3.4.3   Operating Costs (Opex)

Table 9.3 shows OPEX estimation for the years on plateau or peak rates. As this project is tieback with tariff, then total expected OPEX for the field life at low cost (8$/boe) for recoverable gas volume of 300 bcf is 400 MM$ and for high cost (13$/boe) is 650 MM$ for the 4 years of production on plateau period. This total cost must be prorated uniformly to production for the rest of the field life.

For the peak rates on plateau, our yearly Opex would be 50% for 400 MM$/4 totaling 50 MM$ or 50% for 650 MM$/4 totaling 80 MM$ and prorated for the rest of the production years. Again, the Opex distribution during the years is simply indicative and must be based on real operating costs for current and previous projects.

9.3.5   Project Timing and Input to Economic Analysis

A subsea tieback will require roughly 3 years from exploration to production time. During this time period all 2 wells will be drilled (2 wells in one year (2011)). Our project might start up production in 2013. Assuming the exploratory well will be drilled in 2010.

As all the required information are gathered and estimated, net present value (NPV) need to be calculated for the project to evaluate the opportunity considering all the economic parameters. Figure 9.26 summarizes the inputs that any economic package will require, basically the revenues (volumes) and the expenditures.

9.4   Case Study-4: Accumulation with Oil and Gas Reservoir Combined

The offshore discovery is an oil and gas accumulation located on offshore at deepwater (Fig. 9.27). The prospect has an oil initial in place of 500 MMbbl of oil, with low dissolved gas, and 500 bcf of dry gas. The reservoir is Pliocene and Middle Miocene

Date	Oil Production rate stb/day	Gas Production rate MMscf/day	Exploration Drilling MM$ min-max	Appraisal Drilling MM$ min-max	Development Drilling MM$ min-max	Capex MM$ Facility	Capex MM$ Subsea	Capex MM$ pipeline	Abandonment MM$	Opex MM$	Comments
2010			35-64								1 Exploratory
2011					(63-115) × 2		34	36.23			2 wells
2012							34	6.4			
2013		60.00								25-40	
2014		120.00								50-80	
2015		120.00								50-80	
2016		120.00								50-80	
2017		120.00								25-40	
2018		100.00								25-40	
2019		75.00								12.5-20	
2020		51.00								12.5-20	
2021		30.00								12.5-20	
2022		15.00								12.5-20	
2023		6.00									
2024		3.00									
2025											
2026											
2027											
2028											
2029											
2030											
2031											
2032											
Total			35-64		126-230		68	42.63			

Fig. 9.26 Economic analysis inputs

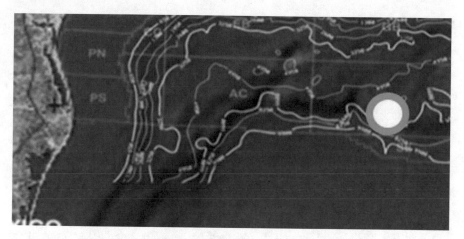

Fig. 9.27 Location of a dry gas prospect in deepwater

Table 9.7 Prospect input
data available

Parameter	Value
OIIP (MMbbl)	500
GIIP (bcf)	500
Water depth (ft)	5,500
Deposition age	Pliocene
Area (acres)	3420
Net pay Thickness (ft)	400
Total depth (ft)	25,000

age Net pay reservoir thickness is 300 ft for oil zone and 100 ft for gas zone. No exploratory wells are available. Table 9.7 shows the input data available from the prospect.

9.4.1   Primary Estimation of the Recovery Factor

The first step to start up the assessment is to predict the recover factor for the oil and gas in place. Using the analogues in Fig. 9.2, the expect recovery factor for oil is about 25% and for gas is around 65% at depth of 25,000 ft. Accordingly, the recoverable volume from the initial oil and gas in place will be 125 MM bbl of oil and 325 bcf of gas.

9.4.2   Rock/Fluid Properties and Well Count Estimation

Now, the number of wells and their initial flow rates must be determined. Therefore, some reservoir rock and fluid properties must to be estimated.

Figure 9.12 shows porosity variation with depth at different deposition environment. At reservoir depth 25,000 ft the estimated average porosity value will be 10–20% for Oligocene (Upper Paleogene). By using this porosity, permeability can be estimated from Fig. 9.4 which is within the range of 5 mD to 75 mD.

Up to now, recovery factor, porosity and permeability ranges using analog information already determined. Most of the Paleogene reservoirs have more or less viscosity equal to 10 cP and API varying from 20 to 22 as proposed in Fig. 9.5.

At low permeability range, KH = 55 × 300/10 = 5150 mD ft/cP, and at high permeability, KH = 575 × 300/10 = 52,250 mD ft./cP.

Using Fig. 9.23, the UR/well would then be between 2 and 10 MMbbl of oil leading to a low case of 125/2, or 63 wells and a high case with 13 wells. The low case with 63 wells would be uneconomical. Therefore, 13 wells scenario will be used.

9.4.3   Well Initial Rates and Notional Production Profile

Figure 9.24 was used to estimate the oil rate per well. This figure is derived from analogues (oil fields with API averaging ranges between 22 API and 24 API). To apply it for gas, we need to assume a viscosity for gas of 0.1 cP and that the rates will be in BOE (barrels of oil equivalent).

At low permeability range, KH = 5100 × 200 = 52,000 mD ft, the average oil rate is about 2,200 bbl/d equivalent to 2 MMscf/d. And at high permeability, KH = 5100 × 200 = 52,000 mD ft, the average oil rate is about 12,000 bbl/d equivalent to 72 MMscf/d. In this project, 10,000 bbl/d equivalent to 60 MMscf/d was proposed as average flow rate for the dry gas play. Therefore, the total gas production from the 2 wells at 60 MMscf/d will be 120 MMscf/d.

Figure 9.24 is derived from analogs (oil fields with API averaging about 24 API). at KH = 522,500, flow rate is around 2200 bbl/d. Typically, the lighter oils would produce higher rates.

Therefore, 13 wells at 2200 bbl/d would yield 28,600 bbl/d. Now we need to generate a production curve that produces a plateau of 28,600 bbl/d and total reserves of 125 MMbbl. Figure 9.28 show the production forecasting scenario.

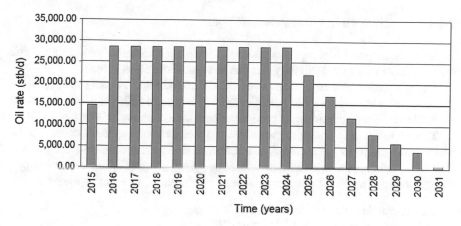

Fig. 9.28 Production forecasting profile

9.4.4 Estimation the Capex and Opex Costs

To execute the development plan, it is required to know the total project cost to drill the 13 development wells, wells completions, surface facilities, evacuation system along with the expenditure costs.

9.4.4.1 Drilling Costs

Expenses will rely on the well depth and the cost of the rig per day. Also, the daily rig rate is varying based on to the rig type, water depth, distance from onshore and target depth. Typically, for onshore, the cost is 100,000$/day, and for deepwater offshore, it can be more higher varying from 600,000 to 800,000$/day (based on 2010 prices). Besides, the total number of drilling days will be a function of depth. Normally, at depth around 20,000 ft, it takes 70 to 80 days, and for deeper wells about 32,000 ft it might reach up to 150 days.

Table 9.2 shows the drilling costs as a function of daily rig rate and drilling time to estimate the dry hole cost.

The range of the well cost shown in the table is the cost of the exploratory well from the analogues. For the appraisal and development wells, we will need to add some additional costs for example, the evaluation and completion costs. As the total volume is 125 MMbbl of oil, we need one exploratory well and 1 appraisal well, and 13 development wells.

The cost of one exploratory well is varying between 55 to 88 MM$, and the cost of one appraisal well assumed to be same as exploratory well plus the well test cost. So, the total appraisal well cost might be within the range between 62.5 to 100 MM$.

So, a development well will cost from 55 MM$ to 88 MM$ plus completion costs (180%) which is between 99 MM$ to 158.4 MM$.

9.4.4.2   Facilities and Subsea Costs (Capex)

Figure 9.17 displays that topsides and hull for semi-submersible production units, they are function of total capacity and can be considered almost linear.

The production unit to be sized to produce 30,000 bbl/d. By using the Fig. 9.17, the hull cost is around 400 MM$ and the top side is around 552 MM$, making a total of 958 MM$. Some contingency costs must be added at about 25% or 1278 MM$.

The next step is to compute the subsea costs and export pipelines costs. For 13 wells, two manifolds are required, one manifold tied with 6 wells and another one tied with 7 wells, plus inter field flow lines and risers. Normally, the costs obtained from a previous project. Some factors need to be assumed to estimate the costs:

Well templates and manifolds 34 MM$/well
SCR production riser at 0.35 MM$/(1000 ft).

Flow lines and export pipelines at 0.23 and 0.16 MM$/(miles). Let's assume that the flow lines length is 55 miles and for export pipeline length equals 30 miles.

These are estimated factors as provided by suppliers. Templates and manifolds cost at 34 MM$/well for 13 wells is 442 MM$, and for production risers at water depth 5,500 ft is 924 MM$. Export riser at 0.28 MM$/(1000 ft) is 15 MM$. Flow line and export pipelines for 5 and 30 miles at 0.16 MM$/(1000 ft) is 48 MM$.

9.4.4.3   Operating Costs (Opex)

The operating costs are more difficult to determine using only charts or factors. They are typically reliant on the peak of the production rate and pipeline tariff rates paid to export production through third party pipelines.

Table 9.3 shows OPEX estimation for the years on plateau or peak rates. As this project is standalone with tariff, then total anticipated OPEX for the field life at as a minimum (12$/boe) of 125 MMbbl and as a maximum (25$/boe) of 125 MMbbl is equaling 1,500 MM$ and 3,125 MM$ respectively for the 9 years of production on plateau. This total cost should be prorated proportionally to production for the remaining years.

For the peak rates on plateau, our yearly OPEX would be 1,500 MM$/9 totaling 50 MM$ or 50% for 3,125 MM$/9 and prorated for the rest of the production years. A minimum of 25% OPEX is suggested for the years that have the production is lower than 25% of peak to take into consideration the maintenance costs and any other costs that are not a direct related to the production.

9.4.5 *Project Timing and Input to Economic Analysis*

We have now calculated most of the input we need for our economic analysis and to the end of our volume to value process (Fig. 9.29).

The timing of the project is just as central as all the other inputs, but it is also flexible reliant how aggressively the company needs to drill the wells and capitalize in advance of the no fracturing of facilities.

Preferably, the company need to drill at least one well per year and consider a period of time to manufacture the facilities and have them in location site.

The table shows an summary for the total project time starting from exploration to first production oil. In the best scenario, this time won't be less than 5 years, and for bigger projects, it might extend up to 10 years or more.

9.5 Study-5: Example of Fields Evacuated to a Shared Facility

Make an economic analysis for an oil prospect, with oil volume in place of 1000 MMbbl, and with low dissolved gas, located in a remote region with a water depth of 5500 ft.

The prospect is of Lower Tertiary (Paleogene) age and located at a depth of 25,000 ft and area of 6800 acres. The net reservoir thickness is about 300 ft with no exploratory wells available in the area, and all the obtained data is from seismic.

The oil production will be displaced through a facility that will be shared with another operator in terms of cost and production.

The difference will be how to model the shared facility. Capex and Opex will be expended uniformly to the production pumped to the facility from each operator.

In this case, the surface treatment facilities need to be upgraded to handle the production from the two fields (Fig. 9.30).

9.5.1 *Facilities and Subsea Costs (Capex)*

Figure 9.17 will be used again to provide an estimation of Capex. Increase the capacity of the facility to handle 60 kbbl/d. By using Fig. 9.17, the hull cost will be 450 MM$ at 60 kbbl/d, and top side cost is around 650 MM$ with a total equaling 1100 MM$. Some of contingency costs, 25%, need to be added. Then, the total cost will be 1375 MM$.

Next step, subsea costs and export pipelines costs need to be calculated. For 25 wells, the project required to buy four manifolds (3 manifolds tied with 6 wells and one manifold tied with 7 wells) plus flow lines and risers. The following are some assumed factors which are used in case project:

Date	Oil Production rate stb/day	Gas Production rate MMscf/day	Exploration Drilling MM$ min-max	Appraisal Drilling MM$ min-max	Development Drilling MM$ min-max	Capex MM$ Facility	Capex MM$ Subsea	Capex MM$ pipeline	Abandonment MM$	Opex MM$	Comments
2010			55-88					pipeline			1 Exploratory
2011				62.5-100	(99-158) × 1	255.6					1 Appraisal+1 well
2012					(99-158) × 3	255.6					3 wells
2013					(99-158) × 3	255.6	147.3333	924			3 wells
2014					(99-158) × 3	255.6	147.3333	15.4			3 wells
2015	14,300.00				(99-158) × 3	255.6	147.3333	48		83.33-173.61	
2016	28,600.00									166.66-347.22	
2017	28,600.00									166.66-347.22	
2018	28,600.00									166.66-347.22	
2019	28,600.00									166.66-347.22	
2020	28,600.00									166.66-347.22	
2021	28,600.00									166.66-347.22	
2022	28,600.00									166.66-347.22	
2023	28,600.00									166.66-347.22	
2024	28,600.00									166.66-347.22	
2025	22,600.00									83.33-173.61	
2026	17,000.00									41.67-86.81	
2027	12,000.00									41.67-86.81	
2028	8,000.00									41.67-86.81	
2029	6,000.00									41.67-86.81	
2030	4,000.00									41.67-86.81	
2031	1,000.00								127.60	41.67-86.81	10% of Total Facility Capex
Total	volume = 125 MMbbl		55-88	62.5-100	1287-2054	1276	442				

Fig. 9.29 Economic analysis inputs

Fig. 9.30 Shared facility
with different projects

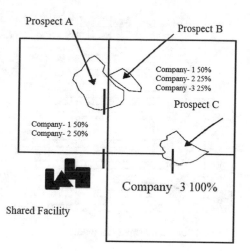

- Well templates and manifolds cost are 34 MM$/well,
- SCR production riser 2 risers per manifold, 8 risers at 0.35 MM$/(1000 ft),
- SCR export riser 1 at 0.28 MM$/(1000 ft),
- Flow lines and export pipelines at 0.23 and 0.16 MM$/(3 miles). Let's assume flow lines extent 5 miles and export pipeline extent 30 miles (let's keep to the same as for case study-1 assuming the flow lines can handle the increase of rate up to 60 kbbl/d).

Well templates and manifolds cost for 25 wells is 850 MM$, and production risers cost is 1,848 MM$. The export riser equals cost is 1540 MM$, and the flow line plus export pipelines cost is 48 MM$.

9.5.2   Operating Costs (Opex)

We will use Table 9.3 to estimate Opex numbers per barrels of oil equivalent for the years on plateau rates.

Therefore, for this field (standalone with tariff), total OPEX need to be estimated for the field life as a minimum (12$/boe) of 250 MMbbl and as a maximum (25$/boe) of 250 MMbbl of totaling 3000 MM$ and 6250 MM$ respectively for the 9 years of production on plateau (till 2014). Again, the total cost must be prorated uniformly to production for the rest of the years.

For plateau rates, our yearly Opex would be 3,000/9 or 6,250/9 and prorated for the rest of the years. A minimum of 25% OPEX is suggested for the years where the production is lesser than 25% of plateau to take into consideration the maintenance costs and other costs which are not a direct related to the production.

Date	Oil Production rate Stb/day	Gas Production rate MMScf/day	Exploration Drilling MM$ min-max	Appraisal Drilling MM$ min-max	Development Drilling MM$ min-max	Capex MM$ Facility	Capex MM$ Subsea	Capex MM$ pipeline	Abandonment MM$	Opex MM$	Comments
2010			55-88								1 Exploratory
2011				62.5-100		275					2 Appraisal+1 well
2012				62.5-100	(99-158) × 1	275					6 wells (2 rigs)
2013					(99-158) × 6	275	283.33	1848			6 wells
2014					(99-158) × 6	275	288.33	15.5			6 wells
2015					(99-158) × 6	275	288.33	48			6 wells
2016	30,000.00				(99-158) × 6					166.50	
2017	60,000.00									333.00	Opex max with Tariff
2018	60,000.00									333.00	
2019	60,000.00									333.00	
2020	60,000.00									333.00	
2021	60,000.00									333.00	
2022	60,000.00									333.00	
2023	60,000.00									333.00	
2024	60,000.00									333.00	
2025	60,000.00									333.00	
2026	48,500.00									166.50	
2027	37,000.00									166.50	
2028	27,000.00									166.50	
2029	19,000.00									83.25	
2030	12,000.00									83.25	
2031	7,500.00								137.50	83.25	
2032	3,700.00										10% of Total Facility Capex
Total			55-88	125-200	2475-3950	1375	850	1911.5			

Fig. 9.31 Economic analysis inputs

9.5.3   Project Timing and Input to Economic Analysis

We have now calculated most of the input we need for our economic analysis and to the end of our volume to value process (Fig. 9.31). The timing of the project (first oil) will be longer than 1 year than in case study-1 because we will need an extra appraisal well as the volume is now two times larger. This second appraisal well can be drilled in the second or third year shifting all the development wells one year later and also deferring the startup of first oil date. Besides, assume that the additional development wells will be drilled concurrently by another rig.

Summary

Today oil and gas investors are fronting several types of risks and uncertainties what makes execution of projects progressively complex. The important goal that investors are trying to obtain is to produce the hydrocarbons in effective and cost efficient manner. This can be reached by confirming that reservoir performance is improved, production is optimized and project risks are reduced.

One of the challenges posed for the oil and gas business is the cost of projects, in terms of the number of wells need to be drilled, the estimation of the amount of hydrocarbon in place and surface and subsurface facility required, etc. Unfortunately, this challenge is existing due to the lack of information and analogues data to make initial evaluation of any new discoveries.

The book was written in a smooth and logical link between one chapter and another to achieve the objective of the book. Nine chapters were inscribed to cover most of the technical information and evaluation required to assess any oil and gas discoveries. The book provides step by step technical details and explanations for every single field development plan stage such as hydrocarbon exploration, reserve estimation, production licensing and field development plan road map. It discussed also the estimation of reservoir rock and fluid properties, along with petroleum reservoir analogues from different prospects. Furthermore, the chapters discussed wells and production functions, well spacing, and optimization number of development wells. The selection of the upstream facilities and subsea production systems were explained in detail together with CAPEX and OPEX costs considerations for oil and gas field development plan process including project timing and input to economic analysis. The last chapter presents the evaluations of five real case studies, which can be used as a good guidance and references to do any similar evaluation for any oil and gas discoveries.

© The Author(s), under exclusive license to Springer Nature Switzerland AG 2020   91
T. A. O. Ganat, *Technical Guidance for Petroleum Exploration and Production Plans*,
SpringerBriefs in Applied Sciences and Technology,
https://doi.org/10.1007/978-3-030-45250-6

References

1. J. Alexander, Advances in reservoir geology, a discussion on the use of analogs for reservoir geology, in *Geological Society*, ed. by M. Ashton (London, Special Publications) 69, pp. 175–194 (1993)
2. G.I. Atwater, E.E. Miller, The effect of decrease in porosity with depth on future development of oil and gas reserves in South Louisiana. Am. Ass. Petrol. Geol. Bull. **49**, 334 (1965)
3. K. Bjørlykke, T. Nedkvitne, M. Ramm, G.C. Salgal, Diagenetic processes in the Brent Group (Middle Jurassic) reservoirs of the North Sea—an Overview, in *Geology of the Brent Group* ed. by A.C. Morton, R.S. Hazeldine, M.R. Giles, S. Brown. Geol. Soc. London, Spec. Publ. 61 (1992), pp. 265–289
4. K. Bjørlykke, M. Ramm, G.C. Samal, Sandstone diagenesis and porosity modification during basin evolution. Geol. Rundschau **78**, 243–268 (1989)
5. S. Bloch, Empirical prediction of porosity and permeability in sandstones. Am. Ass. Petrol. Geol. Bull. **75**, 1145–1160 (1991)
6. R.C. Craze, S.E. Buckley, *A factual analysis of the effect of well spacing on oil recovery* (Drill. Prod. Prac, API, 1949), pp. 144–159
7. T. Dreyer, North sea oil and gas reservoirs—II, Sand body dimensions and infill sequences of stable, humid-climate delta plain channels, ed. by A.T. Buller, E. Berg, O. Hjelmeland, J. Kleppe, O. Torsæter, J.O. Aasen (Graham & Trotman, London, 1990), pp. 337–351
8. DNV GL Oil & Gas, Subsea Facilities- Technology developments, incidents and future trends (2004)
9. S.N. Ehrenber, Relationship between diagenesis and reservoir quality in sandstones of the Garn Formation, Haltenbanken, Mid-Norwegian Continental Shelf. Am. Ass. Petrol. Geol. Bulletin **74**, 1538–1558 (1990)
10. S.N. Ehrenberg, Preservation of anomalous high porosity in deeply buried sandstones by grain-coating chlorite: examples from the Norwegian continental shelf (1993)
11. Energy and Capital. https://www.energyandcapital.com/articles/drilling-the-bakken-formation/73660
12. S.S. Flint, I.D. Bryant, The geologic modeling of hydrocarbon reservoirs and outcrop analogs. International Association of Sedimentologists, Special Publications, p. 15 (1993)
13. H. Füchtbauer, Die Sandsteine in der Molasse nördlich der Alpen. Geol. Rundschau **56**, 266–300 (1967)
14. G.M. Grammar, P.M. Harris, G.P. Eberli, Integration of outcrop and modern analogs in reservoir modeling, integration of modern and outcrop analogs in reservoir modeling: overview and examples from the Bahamas, American Association of Petroleum Geologists, Memoirs, ed. by G.M. Grammar, P.M. Harris, G.P. Eberli, 80, pp. 1–22 (2004)

© The Author(s), under exclusive license to Springer Nature Switzerland AG 2020
T. A. O. Ganat, *Technical Guidance for Petroleum Exploration and Production Plans*,
SpringerBriefs in Applied Sciences and Technology,
https://doi.org/10.1007/978-3-030-45250-6

15. H.H. Haldorsen, Reservoir characterization, simulator parameter assignment and the problem of scale in reservoir engineering, ed. by L.W. Lake, H.B.Carroll Jr.. (Academic Press, New York), pp. 293–340 (1986)

16. R.G. Loucks, M.M. Dodge, L.S. Land, Regional controls on diagenesis and reservoir quality in Lower Tertiary sandstones along the Texas Gulf Coast, in *Clastie Diagenesis* (ed. by D.A. McDonald, R.C. Urdam), AM. Assoc. PETROL. GEOL. MEM., **37**, pp. 15–45 (1984)

17. Libyanlines, (2012). https://libyanlines.wordpress.com/2017/05/11/libya-noc-statement-on-status-of-nc-96-and-nc-97-concessions-wintershall/

18. M. Muskat, Physical Principles of Oil Production. (McGrawHill Book Company, Inc., 1949), pp. 810–904

19. J.C. Maxwell, Influence of depth, temperature and geologic age on porosity of quartzose sandstone. Am. Assoc. Petrol. Geol. Bull. **5**, 697–709 (1964)

20. On the Wight, (2018). http://wig.ht/2g0m

21. Quality stauts report, (2010). https://qsr2010.ospar.org/en/ch07_01.html

22. A. Ryseth, M. Ramm, Prediction of reservoir quality in the Statfjord Formation, Northern Viking Graben: a function of depositional architecture and burial diagenesis. Petrol. Geosci. (submitted) (1995)

23. M. Ramm, Porosity depth trends in reservoir sandstones; Theoretical models related to Jurassic sandstones offshore Norway. Mar. Petrol. Geol. **9**, 553–567 (1992)

24. Shell Exploration and Production-GOM Operations, SEPCO. www.shell.com

25. R.C. Selley, Porosity gradients in the North Sea oilbearing sandstones. J. Geol. Soc. London **135**, 11–132 (1978)

26. M. Scherer, Parameters influencing porosity in sandstones: a model for sandstone porosity prediction. Am. Ass. Petrol. Geol. Bull. **71**, 485–491 (1987). http://dx.doi.org/10.1306/703C80FB-1707-11D7-8645000102C1865D01491423

27. M.B. Standing, D.L. Katz, Density of natural gases. Trans AIME **146**, 140–149 (1942)

28. H. Tokunaga, B.R. Hise, A Method to Determine Optimum Well Spacing, SPE 1673 (1966)

29. H. Tokunaga, B.R. Hise, A method to determine optimum well spacing. Proceedings of the society of petroleum engineers california regional meeting, Santa Barbara, 17–18 November 1966, 7 p. https://doi.org/10.2118/1673-MS (1966)

30. M. Vazquez, H.D. Beggs, Correlations for fluid physical roperty prediction. J. Pet. Technol. **32** (6), 968–970. SPE-6719- PA. https://doi.org/10.2118/6719-PA (1980)

31. K.J. Weber, Reservoir Characterization, How heterogeneity affects oil recovery, ed. by L.W. Lake, H.B. Carroll Jr. (Academic Press, New York, 1986), pp. 487–544

Printed in the United States
By Bookmasters